# MECHANISMS AND SPECIFICITY OF HIV ENTRY INTO HOST CELLS

# ADVANCES IN EXPERIMENTAL MEDICINE AND BIOLOGY



---

## Recent Volumes in this Series

Volume 292
MECHANISMS OF LYMPHOCYTE ACTIVATION AND IMMUNE REGULATION III: Developmental Biology of Lymphocytes
Edited by Sudhir Gupta, William E. Paul, Max D. Cooper, and Ellen V. Rothenberg

Volume 293
MOLECULAR BIOLOGY AND PHYSIOLOGY OF INSULIN AND INSULIN-LIKE GROWTH FACTORS
Edited by Mohan K. Raizada and Derek LeRoith

Volume 294
KYNURENINE AND SEROTONIN PATHWAYS: Progress in Tryptophan Research
Edited by Robert Schwarcz, Simon N. Young, and Raymond R. Brown

Volume 295
THE BASAL FOREBRAIN: Anatomy to Function
Edited by T. Celeste Napier, Peter W. Kalivas, and Israel Hanin

Volume 296
PLASTICITY AND REGENERATION OF THE NERVOUS SYSTEM
Edited by Paola S. Timiras, Alain Privat, Ezio Giacobini, Jean Lauder, and Antonia Vernadakis

Volume 297
NEW ASPECTS OF HUMAN POLYMORPHONUCLEAR LEUKOCYTES
Edited by W. H. Hörl and P. J. Schollmeyer

Volume 298
SENSORY NERVES AND NEUROPEPTIDES IN GASTROENTEROLOGY: From Basic Science to Clinical Perspectives
Edited by Marcello Costa, Calogero Surrenti, Sergio Gorini, Carlo Alberto Maggi, and Alberto Meli

Volume 299
ADVANCES IN PERINATAL THYROIDOLOGY
Edited by Barry B. Bercu and Dorothy I. Shulman

Volume 300
MECHANISMS AND SPECIFICITY OF HIV ENTRY INTO HOST CELLS
Edited by Nejat Düzgüneş

---

A Continuation Order Plan is available for this series. A continuation order will bring delivery of each new volume immediately upon publication. Volumes are billed only upon actual shipment. For further information please contact the publisher.

# MECHANISMS AND SPECIFICITY OF HIV ENTRY INTO HOST CELLS

Edited by

**Nejat Düzgüneş**

University of the Pacific
San Francisco, California

PLENUM PRESS • NEW YORK AND LONDON

**Library of Congress Cataloging-in-Publication Data**

---

Workshop on Mechanisms and Specificity of HIV Entry into Host Cells (1989 : San Francisco, Calif.)
Mechanisms and specificity of HIV entry into host cells / edited by Nejat Düzgüneş.
p. cm. -- (Advances in experimental medicine and biology ; v. 300)
"Proceedings of a Workshop on Mechanisms and Specificity of HIV entry into Host Cells, held June 19-20, 1989, in San Francisco, California"--T.p. verso.
Includes bibliographical references and index.

1. HIV (Viruses)--Congresses. 2. Host-virus relationships--Congresses. 3. AIDS (Disease)--Pathogenesis--Congresses. I. Düzgüneş, Nejat. II. Title. III. Series.
[DNLM: 1. Antigens, CD4--physiology--congresses. 2. HIV--physiology--congresses. 3. Viral Proteins--metabolism--congresses. 4. Virus Replication--physiology--congresses. W1 AD559 v. 300 / QW 166 W926m 1989]
QR414.6.H58W67 1989
616'.0194--dc20
DNLM/DLC
for Library of Congress 91-24096
CIP

---

Proceedings of a Workshop on Mechanisms and Specificity of
HIV Entry into Host Cells,
held June 19–20, 1989, in San Francisco, California

**ISBN-13: 978-1-4684-5978-4 e-ISBN-13: 978-1-4684-5976-0**
**DOI: 10.1007/978-1-4684-5976-0**


**Softcover reprint of the hardcover 1st edition 1991**

A Division of Plenum Publishing Corporation
233 Spring Street, New York, N.Y. 10013

# PREFACE

The first reports on the tropism of the human immunodeficiency virus (HIV), then called human T-cell lymphotrophic virus type III (HTLV-III) or lymphadenopathy associated virus (LAV), indicated an almost absolute specificity of the virus for T4 (helper/inducer) lymphocytes. It became apparent a few years later that macrophages were also infectable by HIV. The common cellular receptor for the virus on these cell types was CD4. A remarkable series of studies by five different groups, four of which were published in the same issue of Nature, showed that recombinant soluble CD4 could inhibit the infectivity of the virus. Subsequent studies began to reveal, however, that cells not expressing CD4 could also be infected, thus raising the possibility that almost any cell in the body could act as a latent reservoir for the virus. Recent observations that the infectivity of patient isolates is not inhibited at the low concentrations of rsCD4 that inhibit laboratory strains, also indicate the importance of CD4-independent mechanisms. The mechanisms of fusion of HIV with cellular membranes are under investigation by a large number of laboratories. These studies range from the determination of the oligomeric structure of the envelope glycoproteins, to the analysis of the function of various regions of the proteins by site-directed mutagenesis, to the fluorimetric monitoring of membrane fusion.

This book originated as the proceedings of a Workshop on the Mechanisms and Specificity of HIV Entry into Host Cells, held at the University of California, San Francisco in June 1989. Thus, several of the chapters are followed by edited transcripts of the discussions following each presentation at the Workshop. The editor hopes that the record of the spontaneous exchange of information during these sessions will help the reader generate new ideas in this field. The chapters were prepared after the Workshop, and may have additional data not presented at the meeting. Complementing these chapters are additional ones by authors who were invited subsequently to participate in the book.

In a field that moves rapidly, the value and timeliness of publishing a book, rather than original research articles, may be questioned. The final judge is, of course, the reader. The editor's view is that it is important to place our knowledge in perspective, to summarize, speculate and re-evaluate, as the authors of this volume have done. The reviews, as well as original data, presented in this book will be useful not only for researchers working in this field, but also for beginners and those researchers outside the field looking in. While journal articles are essential for immediate communication and progress, books have a way of making it to scientists' desks, armchairs and bookshelves long after the journals are shelved in archives. I am therefore indebted to all the authors for their contributions to the volume.

In the production of the book, James Harris of the Cancer Research Institute, UCSF, spent innumerable hours transcribing the discussions and some of the lectures, which provided the framework for some of the chapters. Cynthia Vincent provided invaluable help in editing the transcripts. I am grateful for their contributions. The organization of the workshop benefited immensely from the assistance of Sara Burke of the UCSF Extended Programs in Medical Education. The primary funding for the workshop was provided by the State of California Universitywide Task Force on AIDS (now the "Universitywide AIDS Research Program"). I would like to thank Dr. Paul Volberding for allocating additional funds for the workshop from the UCSF Center for AIDS Research at San Francisco General Hospital. Funds were also provided by the UCSF Extended Programs in Medical Education for the workshop, and the Cancer Research Institute for the production of the proceedings.

Finally, in deep appreciation of all the support they have given me, I dedicate this book to my family.

Nejat Düzgüneş
San Francisco

CONTENTS

# VIRAL AND CELLULAR FACTORS INFLUENCING HIV TROPISM

Jay A. Levy

Department of Medicine and Cancer Research Institute

University of California, School of Medicine

San Francisco, CA, 94143-0128

## Introduction

Studies of the human immunodeficiency virus (HIV) have indicated that virus-infected cells are most likely the major source of infection, and any approach that kills only the virus would not control HIV infection (1). Early observations suggested that the CD4+ T helper lymphocyte was the major cell type infected by HIV (2). Subsequent studies have demonstrated that a wide variety of cells coming from the hematopoietic and central nervous systems, the skin and the bowel are susceptible to HIV (3). Based on observations on a variety of HIV isolates it appears likely that, depending on the particular virus strain infecting the host, any human cell could be infected by HIV (4) (Table 1). Its detection might be difficult in some cells in which viral replication takes place at a low level. Nevertheless, the presence of HIV could be revealed by Southern blot or polymerase chain reaction (PCR) techniques. Most importantly, if tissues in the body are infected by HIV, even if producing low levels of the virus, their function may be compromised.

**Table 1**
**Human Tissues and Cells Found Susceptible to HIV Infection**

Human Tissues

| | | |
|---|---|---|
| Adrenal | Eye | Lung |
| Bone marrow | Heart | Placenta |
| Bowel | Joints | Prostate |
| Brain | Kidney | Skin |
| Cervix | Liver | Testes |
| | | Thymus |

*Mechanisms and Specificity of HIV Entry into Host Cells*
Edited by N. Düzgüneş, Plenum Press, New York, 1991

## Human Cells

| | |
|---|---|
| *Hematopoietic*: | Stem cells, B lymphocytes, T lymphocytes, macrophages, dendritic cells |
| *Skin*: | Langerhans cells, fibroblasts |
| *Bowel*: | Columnar; goblet; enterochromaffin; macrophages and lymphocytes (lamina propria); bowel carcinoma cell lines; hepatoma cell lines |
| *Brain* | Microglia/macrophages, astrocytes and other glial cells; endothelial; glioma cell lines. |
| *Other* | Lung fibroblasts; placental cell lines |

## HIV Replication in Infected CD4+ and CD4- Cells

Some of the initial studies demonstrating that other receptors on cells might be needed for HIV infection focussed on T cells that vary in their expression of the CD4 molecule. Work by Kikkawa, et al (5) with various clones of the MOLT-4 cell line, and studies by Louise Evans in our laboratory (6) using a variety of established T cell lines, indicated that virus infection and replication depended on other parameters besides the extent of expression of CD4. Some T cells, for instance, with high level expression of CD4 were resistant to HIV infection (5,6). Moreover, studies by Maddon and co-workers indicated that even cells (e.g., rodent), induced by molecular techniques to express the CD4 protein to high level could not be infected by HIV (7). All these observations have led to the hypothesis that two steps may be involved in HIV entry into cells (8). First, HIV attaches to the CD4 molecule through its gp120 envelope glycoprotein, and second, it fuses with another cell membrane receptor most likely using the viral fusogenic domain of gp41 (8,9). Thus, if CD4 alone was present on the cell with no putative "fusion receptor", the virus might attach but not enter (e.g., some T cell lines). If both receptors were present, virus infection would be highly efficient. If only the fusion receptor was expressed, the virus might enter the cell but with less frequency since the CD4 protein that helps bring HIV close to the cell membrane is absent. A similar two step mechanism for virus entry has been described for paramyxoviruses (10). The search then is on to find the cell fusion receptor that may be a common denominator for HIV entry into all human cells.

Evidence that CD4- cells are susceptible to HIV came from the early studies of human glioma cell lines. Some of these brain-derived cells were shown to lack CD4 messenger RNA and protein expression and yet could be infected and replicate certain strains of HIV (11). Subsequent studies with other brain-derived cells, human fibroblastoid cells, as well as liver and brain cell lines have substantiated the fact that human cells can be infected even if not expressing CD4 (12-17). These observations along with the finding of HIV in bowel and renal epithelia (18,19) that lack CD4 protein expression support the above mentioned hypothesis that other receptors for HIV entry exist on human cells. The experimental procedures used to demonstrate this CD4-independent infection by HIV are listed in Table 2. These methods include the inability to block the infection with soluble CD4 or with antibodies to the CD4 protein on the cell surface (e.g., Leu3a).

In regard to CD4- cells, Dr. Masatoshi Tateno in our laboratory (16) performed extensive studies on CD4- fibroblastoid cells as well as human skin fibroblasts. He found that some virus isolates from the blood as well as the brain, showing either cytopathic or non-cytopathic properties, could infect and replicate in these fibroblastoid cells. This infection was not blocked by antibodies to CD4. Moreover, human antibodies to the HIV strains used differed in their ability to block infection of T cells vs. fibroblasts by HIV. Clearly distinct regions of the viral envelope appeared involved in the infection of these different cell types. These observations placed further emphasis on the possible presence of other cellular receptors for HIV.

**Table 2**
**Evidence That CD4 Is Not Always Involved in Infection of Some Human Cells.**

1. Antibodies to the CD4 molecule (e.g., Leu3a) do not block HIV infection.

2. Soluble CD4 does not block HIV infection.

3. Antibodies that neutralize HIV infection of CD4+ T lymphocytes may not neutralize virus infection of CD4- fibroblasts.

4. CMV infection of CD4- fibroblasts induces an Fc receptor. Antibody-mediated enhancement of these cells can take place via the Fc receptor.

5. Human T cells with high level expression of CD4 cannot be infected by certain HIV strains.

6. CD4+ macrophages can not be infected by some HIV-1 strains.

Generally, detecting HIV infection of CD4- cells is difficult because of the low level replication of the virus. The studies require controls indicating elimination of input virus. Dr. Shenbei Tang in our laboratory showed that a very short exposure of HIV to standard preparations of trypsin (40-100 ug/ml) quickly inactivates the agent. Thus, to exclude the possibility of input virus remaining after several washings of the CD4- cells, trypsinization procedures should be performed following the initial infection of the cells. One other important observation made with CD4- cells was the requirement for cocultivation of these cells with peripheral blood mononuclear cells (PMC) of seronegative donors (11,14,16,17). This procedure reveals the low levels of virus replication by producing a rapid peak of HIV production in the PMC culture. This observation suggests that even if fibroblasts were infected in the host, and these cells replicated HIV to low titers, circulating PMC could pick up the HIV and spread the infection throughout the body.

## Intracellular Control of HIV Replication

One other important observation made in these studies of host cell susceptibility to HIV was the differences in virus replication occurring after HIV entry. Dr. Louise Evans demonstrated this biologic feature with a variety of human T cell lines, and also with PMC from individuals of different ethnic backgrounds (6). The varying levels of virus replication particularly in the PMC from different donors demonstrated dramatically the role of intracellular controls for virus replication that must be appreciated in all studies of HIV tropism. Other studies have shown that cytokines present in the culture medium (conceivably human plasma *in vivo*) can enhance HIV replication most likely by inducing cellular proteins that interact with the viral LTR (20,21). In addition, the role of HIV regulatory genes in determining this replicative ability, once viral entry has taken place, has been examined. The addition of *tat*-expressing vectors to cells relatively resistant to replication by some HIV strains makes these cells now susceptible to high level virus replication (22). Most likely in these studies the determining factor in host range was the specific viral *tat* protein. In other studies, the inhibitory effect of the *nef* protein on virus replication has been observed and seems to be best explained by an interaction with cellular factors (see below). All these observations on intracellular regulation of HIV replication, however, require further evaluation.

## Organ Systems Affected By HIV Infection and Tropism

HIV has been detected directly in cells of the brain (23,24) and in the bowel and kidney epithelia (18,19). Other reports on HIV in the heart and lung have also appeared (25,26). In the brain, its presence has been noted in microglia and macrophages, as well astrocytes and other glial cells (23,24). In the bowel, it has been found in the crypt cells, including columnar and goblet cells, and the hormone-producing enterochromaffin cells (18). This latter cell type, once believed to be derived from the neural crest, most likely comes from the anlage of the gut and shows a predilection for infection by HIV. HIV infection of enterochromaffin cells in the gastrointestinal tract could explain the disorders of motility and absorption observed in some patients. Although Adachi et al (27) found that some transformed bowel cell lines express CD4, it is generally believed that the bowel epithelium does not, and this tissue is not infected by HIV through the CD4 molecule.

Our studies on brain infection by HIV began in early 1984, when we were asked to determine if a 32 year old homosexual man who died with dementia and seizures, was infected by the AIDS virus. Because his helper/suppressor ratio was near normal (1.4) it was uncertain whether he was infected with the virus. At autopsy, we were able to recover the AIDS retrovirus from lymph nodes and from sections of the brain (28). This observation was one of the first to indicate direct HIV infection of the central nervous system (CNS). It demonstrated as well that neurologic AIDS might be present in individuals with very little evidence of immunologic disorder as revealed by the helper/suppressor ratio. The portion of the CNS showing the greatest pathology was the brain stem in which vacuolar myelopathy was noted. Working with Jay Nelson and Clayton Wiley, we found, using *in situ* hybridization and histochemistry, that macrophages, glial cells, and stellate-shaped cells in the brain had evidence of HIV infection (28). Although monoclonal antibodies specific for oligodendrocytes were not available, the morphology of the infected stellate-shaped cells closely resembled this brain-derived glial cell. We conclude that oligodendrocytes can be infected by HIV and that the dementia observed in patients may be due to the poor production by this cell type of myelin that helps in nerve transmission.

Studies on virus isolates from the brain have indicated that they differ from those conventionally obtained from the blood of infected individuals by a variety of properties (Table 3) (29). Most importantly, the brain-derived strains replicate to high levels in macrophages, and although infecting CD4+ blood lymphocytes, are not usually cytopathic for these cells. This latter observations might explain the near normal helper/suppressor ratios in some individuals presenting with neurologic signs. Furthermore the brain isolates were not as neutralizable by serum as were blood isolates. These findings on different biologic and serologic properties have been supported as well with brain and blood isolates recovered from the same individual (29).

**Table 3**

**Biologic and Serologic Properties of HIV Strains Coming from the Blood, Brain and Bowel**

| Properties | HIV Strains From: Blood | Brain | Bowel |
|---|---|---|---|
| Growth in CD4+ lymphocytes | ++ | ++ | ++ |
| Growth in macrophages | ± | ++ | + |
| Growth in established cell lines | ++ | - | ± |
| Cell killing | + | - | + |
| Serum neutralization | ++ | - | + |

Furthermore, work undertaken in our laboratory on blood and bowel isolates are also showing some differences in their features that may indicate a tropism selective for the gastrointestinal tract (Table 3) (30). From all these observations one could conclude that variants of HIV strains might evolve in the host over time and take on biologic properties that permit them to infect and cause pathology in particular tissue systems such as the brain or the bowel. Thus, identifying the regions of the HIV genome responsible for this tropism could have important relevance to approaches at controlling pathogenesis.

One other noteworthy finding on cell tropism has been made by Cecilia Cheng-Mayer in our laboratory (31). HIV strains grown in established T cell lines were found to change their tropism so that they replicated with faster kinetics and to higher titer in a variety of other cell types. The only modification in the virus correlating with this increased cellular tropism was the glycosylation pattern of gp120. While the evidence is not conclusive that this post-infection modification of the HIV envelope is cell-dependent, the data thus far strongly suggest glycosylation itself can change the tropism of HIV strains. Selection of a virus variant by passage of HIV in different cell types is unlikely in these studies, since we obtained the same results with a molecularly cloned HIV strain, HIV-$1_{SF2mc}$, passed through a variety of cell types (31).

When HIV-$1_{SF2mc}$ was used to infect the PMC from different individuals, the host range was also modified. The virus progeny from some cells grew in certain established cell lines: some in others (31). These observations suggest that one virus infecting different individuals might show a different tropism for cell types in the host depending on the post-infection modification (e.g., glycosylation) of the virus occurring within the cells of the infected individual. This parameter would add to the other observation, cited above, that the replicative ability of HIV can be influenced by the intracellular milieu of PMC from different individuals (6). All these findings suggest that if one HIV strain infects two different individuals, the relative growth of that strain in the PMC of each individual, and the particular envelope proteins made in these cells might influence its spread and tissue tropism in the host and thus its pathogenesis.

## Molecular Basis of Viral Tropism

The properties of viral host range have been linked to genetic sequences using molecular DNA clones of HIV strains with different biologic properties. In our laboratory, for example, we have generated clones of the HIV-1 strains, SF2, SF13, SF128A (a brain isolate), and SF33 (a highly cytopathic strain). The studies completed thus far indicate that the viral envelope region is responsible for cell tropism, CD4 down-modulation, and cytopathicity (32,33). Recent molecular studies further pinpoint a region in the C-terminal portion of the HIV envelope that determines macrophage and T cell tropisms (34). This domain includes part of the V3 region, the principal neutralizing region of HIV-1 (35). These new findings could be helpful in antiviral approaches in which the regions on the viral surface required for HIV entry might be modified.

## Virus Entry by Enhancing Antibodies

In the course of studies examining neutralization of HIV strains, Jacques Homsy in our group found that numerous immunizations of guinea pigs with selected viral isolates did not lead to neutralizing antibodies to the particular viral strain used. Instead, antibodies were produced that enhanced virus replication 2-4 fold (36). In subsequent studies of HIV-infected chimpanzees with no neutralizing antibodies, enhancing antibodies were also noted. Finally, certain human sera were found to enhance virus replication in CD4+ lymphocytes as well as macrophages. One serum of interest had neutralizing antibodies to two HIV-1 strains, but enhanced infection by a third strain (36). The latter one had come from Africa (SF170). These findings have important implications for vaccine development. Immunization against one HIV strain might give protection against that isolate, but enhance infection by another. The epitope(s) responsible for enhancement should therefore be defined before individuals are immunized with purified viral proteins.

These studies have also indicated that the phenomenon of antibody-dependent enhancement (ADE) can play an important role in cell tropism. Certain HIV-1 and HIV-2 strains that are not monocyte-tropic can be brought into macrophages after reaction with antibodies that cause enhancement of virus infection (37). These antibodies can also permit the strains to infect human fibroblasts (37) that previously were resistant to infection. The entry site of HIV by ADE in macrophages is the receptor (37), but for T cells or fibrobasts the FcRIII region has not been identified. It is noteworthy that Homsy et al (37) do not find CD4 plays a role in this ADE of HIV infection of macrophages. The Fc receptor alone can function as the sole entry point for HIV:antibody complexes. This conclusion, however, is controversial, since work with cell lines by other investigators have suggested a role for CD4 (38). Ed Robinson and colleagues have described HIV enhancement in which the complement receptor is involved (39-41). They have demonstrated that a large number of sera in the presence of complement will enhance HIV infection of a CD4+ T cell line; in this case the CD4 molecule also seems necessary for infection (41).

Two conclusions can be reached from all these observations on ADE. 1) Enhancing antibodies do exist in HIV infection as they have been described with dengue virus and other viral diseases (42). The enhancement need not depend on the CD4 receptor and thus would not be blocked by soluble CD4 therapy. This point, made by Homsy et al (37), has been dramatically shown by the infection of CD4- fibroblasts with cytomegalovirus. This virus, like other herpes viruses, induces an Fc receptor in the infected cells (43). Subsequently, HIV:antibody complexes infect the CD4- fibroblasts by this receptor. 2) Enhancement does not appear to be a laboratory artifact, since individuals with varying antibody profiles have been studied over time as they advance from an asymptomatic state to disease. During this interval, the antibodies change from the neutralizing to enhancing type for the particular strain emerging during the course of infection (44). The virus appears to change and take advantage of the immune response to increase its spread. These observations on ADE have suggested therapeutic approaches in which a hybrid molecule of CD4 and CD16 (FcRIII) might be used to inhibit HIV infections.

## Latency

Most of the discussion on cell tropism has focussed on viral attachment, penetration and replication in a variety of different cell types and using different receptors. The viral envelope appears to be important for the initial steps of this infection, and the intracellular milieu certainly contributes to the relative ability of a virus to replicate in the cells infected. The heterogeneity of HIV strains is also reflected in the ability of some HIV isolates to infect and remain latent (or silent) in a cell type. In this state, the virus is integrated in the cellular genome but expresses very little if any viral RNA or protein. We have shown, for example, that an HIV-1 isolate from the Dominican

Republic, SF247, initially obtained from the PMC of a patients of Prof. Ellen Koenig in Santo Domingo, induces latency quite rapidly in the Jurkat T cell line (28). Nevertheless, the same strain grows without any delay in PMC. Thus, selected cell types may be susceptible to the induction of latency by particular HIV strains. When halogenated pyrimidines (e.g., IUDR) or other activators of retrovirus replication are applied to the latently infected cells, virus replication takes place (28). However, often, as is seen with animal retroviruses, the virus may return to its latent state in 3-4 days (28).

The phenomenon of latency, first demonstrated with HIV-1 by Folks, Hoxie and their co-workers (45,46) can be observed in CD4+ lymphocytes and because of virus differences appears related to particular HIV gene products. Early studies in our laboratory demonstrated that the elimination of the *nef* gene from a molecular clone of HIV-1 led to the generation of a virus that replicated much more rapidly and had a wider host range than the initial wild type HIV-1 (47). Subsequent studies in which the *nef* gene was introduced into established T cell lines, have recently shown that production of the p27 *nef* gene product can reduce or prevent replication of both HIV-1 and HIV-2 strains (48). Nevertheless, these findings have revealed that viral heterogeneity is evident in the response to the *nef* gene. Viruses that are highly cytopathic, occurring when the individual progresses to disease, are much less susceptible to the effects of *nef*. This observation, therefore, can explain the inability of some laboratories to show a "silencing" effect of *nef* on HIV replication (49,50). This difference in susceptibility to *nef* was particularly evident when a pair of isolates from the same individual (HIV-$1_{SF2}$ and HIV-$1_{SF13}$) was used to infect HUT 78 T cells expressing *nef*. Only the SF2 strain was suppressed in its replication. SF13, a highly cytopathic strain isolated from the individual a few months before he died from AIDS, had no sensitivity to *nef* (48). Molecular studies using these two strains may help uncover the region of the virus responsive to the *nef* protein.

Guy and associates demonstrated that *nef* expressed in bacteria has GTP binding and GTPase activity (51). They also reported that *nef* down-modulates CD4. This latter property of this viral gene has not been observed by Cheng-Mayer in most of the studies undertaken in our laboratory with tranfection of *nef* into T cell lines (48). Nevertheless, work by Louise Poulin in our group has confirmed, using *in vitro* translation of *nef*, several of its biochemical properties including GTP binding and kinase activity (52). We are presently examining whether a cellular protein, apparently phosphorylated by *nef*, could be important in the induction of viral latency.

What has become evident from these studies of *nef* is that this viral protein alone cannot induce latency, but needs to interact with certain cellular factors, and these may differ in their ability to control HIV replication. *Nef* expressed in some fibroblastoid cells, for instance, will not suppress HIV replication substantially (Cheng-Mayer et al, unpub. obs.). In time, we hope to capitalize on the *nef* gene to use its ability to inhibit virus replication as a potential antiviral approach.

## Conclusions

Studies in our laboratory and others have indicated the wide variety of cell types that can be infected by HIV-1 and HIV-2 strains. Whereas initially the T helper lymphocyte was considered the major target, studies since 1983 have clearly demonstrated infection of a wide variety of cells and tissues in the body. The CD4 molecule appears to be the major receptor for viral entry but the infection of CD4- cells indicates alternative methods for virus infection. Moreover, the studies of serum enhancement of HIV infection demonstrate that the Fc or complement receptor can also permit virus entry into cells. Furthermore, once virus penetration has occurred, the intracellular milieu can determine to what extent virus replication takes place. This latter observation is highlighted by the different replicating abilities of HIV strains in the same cell lines or in the peripheral white cells of different individuals. Depending on the extent of envelope glycosylation by the cell, the tropism of HIV can also potentially be affected. This influence of intracellular milieu can be observed as well by the probable interaction of cellular factors with viral regulating proteins that can control HIV replication. For example, latent viral infection may result from the expression of *nef* in some cells. Clearly, then, both viral and cellular parameters are important in HIV tropism (Table 4).

**Table 4**
**Viral and Cellular Features Influencing HIV Tropism**

*A.* *Surface Events*

1. CD4 receptor
2. "Putative" fusion receptor
3. Fc receptor for enhancing antibodies
4. Viral envelope region that determines macrophage and T cell tropisms.

*B.* *Intracellular Events*

1. Cellular factors influencing HIV replication
2. Cellular factors affecting envelope glycosylation
3. Cellular factors interacting with the viral LTR
4. Cellular factors interacting with the viral *nef* protein

Certainly, the spread of HIV must play a role in its pathogenesis. Thus, any findings that can lead to therapeutic approaches to limit the wide tropism of the virus should be helpful for preventing the progression of AIDS in infected individuals.

Acknowledgements

The research studies from the authors laboratory were supported by grants from the National Institutes of Health (P01 AI24286, R01 AI25284) and the California State Universitywide Task Force on AIDS. Christine Beglinger is thanked for preparation of the manuscript.

References

1. Levy, J.A. The tansmissison of AIDS: The case of the infected cell. J. Am. Med. Assoc. 259:3037 (1988).

2. Klatzmann D., Barre-Sinoussi F., Nugeyre, M., et al. Selective tropism of lymphadenopathy-associated virus (LAV) for helper-inducer T lymphocytes. Science 225:59 (1984).

3. Castro B., Cheng-Mayer, C., Evans, L.A., and Levy, J.A. HIV heterogeneity and viral pathogenesis. AIDS 2:s17 (1988).

4. Levy, J.A. Changing concepts in HIV infection: Challenges for the 1990s. AIDS 4:1051 (1990).

5. Kikkawa R., Yoyanagi Y., Harada, S., et al. Differential susceptibility to the acquired immunodeficiency syndrome retrovirus in cloned cells of human leukemic T-cell lines Molt-4. J. Virol. 57:1157 (1986).

6. Evans, L.A., McHugh T.M., Stites D.P., and Levy J.A. Differential ability of human immunodeficiency virus isolates to productively infect human cells. J. Immunol. 138:3415 (1987).

7. Maddon P.J., Dalgleish A.G. McDougal J.A., et al. The T4 gene encodes the AIDS virus receptor and is expressed in the immune system and the brain. Cell 47:333 (1986).

8. Levy J.A., Human immunodeficiency viruses and the pathogenesis of AIDS. J. Am. Med. Assoc. 261:2997 (1989).

9. Gallaher W.R. Detection of a fusion peptide sequence in the transmembrane protein of human immunodeficiency virus. Cell 50:327 (1987).

10. Merz D.C., Scheid, A. and Choppin, P.W. Importance of antibodies to the fusion glycoprotein of paramyxoviruses in the prevention of spread of infection. J. Exp. Med. 151:275 (1980).

11. Cheng-Mayer C., Rutka J.T., Rosenblum M.L., McHugh T., Stites D.P, and Levy J.A. The human immunodeficiency virus (HIV) can productively infect cultured human glial cells. Proc. Proc. Natl. Acad. Sci. 84:3526 (1987).

12. Chiodi, F., Fuerstenberg, S., Gidlund, M., Asjo, B., and Fenyo, E.M. Infection of brain-derived cells with the human immunodeficiency virus. J. Virol. 61:1244 (1987).

13. Dewhurst, S., Sakai, K., Bresser, J., Stevenson, M., Evinger-Hodges, M.J. and Volsky, D.J. Persistent productive infection of human glial cells by human immunodeficiency virus (HIV) and by infectious molecular clones of HIV. J. Virol. 61:3774 (1987).

14. Clapham, P., Weber, J.N., Whitby, D., McKintosh, K., Dalgleish, A.G., Maddon, P.J., Deen, K.C., Sweet, R.W. and Weiss, R.A. Soluble CD4 blocks the infectivity of diverse strains of HIV and SIV for T cells and monocytes but not for brain and muscle cells. Nature 337:368 (1989).

15. Harouse J.M., Kunsch C., Hartle H.T. et al. CD4-independent infection of human neural cells by human immunodeficiency virus type 1. J. Virol. 63:2527 (1989).

16. Tateno M., Gonzalez-Scarano F., and Levy, J.A. The human immunodeficiency virus can infect CD4-negative human fibroblastoid cells. Proc. Natl. Acad. Sci. 86:4287, 1989.

17. Cao Y.Z., Friedman-Kien A.E., Huang Y.X., et al. CD4-independent, productive human immunodeficiency virus type 1 infection of hepatoma cell lines in vitro. J. Virol. 64:2553 (1990).

18. Nelson J.A., Wiley C.A., Reynolds-Kohler C, Reese C.E., Margaretten W., and Levy, J.A. Human immunodeficiency virus detected in bowel epithelium from patients with gastrointestinal symptoms. Lancet 1:259 (1988).

19. Cohen A.H., Sun N.C.J., Shapshak P., and Imagawa D.T. Demonstration of human immunodeficiency virus in renal epithelium in HIV-associated nephropathy. Mod. Pathol. 2:125 (1989).

20. Nabel G., and Baltimore D. An inducible transcription factor activates expression of human immunodeficiency virus in T cells. Nature 326:711 (1987).

21. Osborn L., Kunkel S., and Nabel G.J. Tumor necrosis factor alpha and interleukin 1 stimulate the human immunodeficiency virus enhancer by activation of the nuclear factor $_k$B. Proc. Natl. Acad. Sci. 86:2336 (1989).

22. Schwartz S., Felber B.K. Fenyo E.-M., and Pavlakis G.N. Rapidly and slowly replicating human immunodeficiency virus type 1 isolates can be distinguished according to target-cell tropism in T-cell and monocyte cell lines. Proc. Natl. Acad. Sci. 86:7200 (1989).

23. Koenig, S., Gendelman, H.E., Orenstein, J.M., Dal Canto, M.C. et al. Detection of AIDS virus in macrophages in brain tissue from AIDS patients with encephalopathy. Science 233:1089 (1986).

24. Wiley, C.A., Schrier R.D., Nelson J.A., Lampert P.W., and Oldstone M.B.A. Cellular localization of human immunodeficiency virus infection within the brains of acquired immune deficiency syndrome patients. Proc. Natl. Acad. Sci. 83:7089 (1986).

25. Leonard H., Calabrese D.O. Proffitt M.R., et al. Congestive cardiomyopathy and illness related to the acquired immunodeficiency syndrome (AIDS) associated with isolation of retrovirus from myocardium. Ann. Intern. Med. 107:691 (1987).

26. Plata F., Garcia-Pons F., Ryier A., et al. HIV-1 infection of lung alveolar fibroblasts and macrophages in humans. AIDS Res. Hum. Retroviruses 6:979 (1990).

27. Adachi, A., Koenig, S., Gendelman, H.E., Daugherty, D., Gattoni-Celli, S., Fauci, A.S. and Martin, M.A. Productive, persistent infection of human colorectal cell lines with human immunodeficiency virus. J. Virol. 61(1):209 (1987).

28. Levy J.A., Evans L.A., Cheng-Mayer C, et al. The biologic and molecular properties of the AIDS-associated retrovirus that affect antiviral therpay. Ann. Inst. Pasteur 138:101 (1987).

29. Cheng-Mayer C., Weiss C., Seto D., and Levy J.A. Isolates of human immunodeficiency virus type 1 from the brain may constitute a special group of the AIDS virus. Proc. Natl. Acad. Sci. 80:8575 (1989).

30. Barnett S., Barboza A.J., Wilcox C.M., Forsmark C.E., and Levy J.A. Characterization of HIV-1 strains recovered from the bowel. 6th International Conference on AIDS. San Francisco, June 1990 [abstract S.A. 68].

31. Cheng-Mayer C., Seto D., and Levy J.A. Altered host range of HIV-1 after passage through different human cell types. Virology, in press.

32. York-Higgins D, Cheng-Mayer C, Bauer D, Levy J.A., and Dina D. Human immunodeficiency virus type 1 cellular host range, replication and cytopathicity are linked to the envelope region of the viral genome. J. Virol. 64:4016 (1990).

33. Cheng-Mayer C., Quiroga M., Tung, J.W., Dina D.,and Levy J.A. Viral determinants of human immunodeficiency virus type 1 T-cell or macrophage tropism, cytopathogenicity, and CD4 antigen modulation. J. Virol. 64:4390 (1990).

34. Shioda, T., Levy, J.A. and Cheng-Mayer, C. Macrophage and T cell line tropisms of HIV-1 are determined by specific regions of the envelope gp120. Nature (in press).

35. LaRosa G.J., Davide J.P, Weinhold K., et al. Conserved sequence and structural elements in the HIV-1 principal neutralizing determinant. Science 249:932 (1990).

36. Homsy J., Tateno M., and Levy, J.A. Antibody-dependent enhancement of HIV infection. Lancet 1:1285 (1988).

37. Homsy J., Meyer M., Tateno M, Clarkson S., and Levy J.A. The Fc and not the CD4 receptor mediates antibody enhancement of HIV infection in human cells. Science 244:1357 (1989).

38. Takeda, A. and Ennis, F.A. FcR-mediated enhancement of HIV-1 infection by antibody AIDS Research and Human Retroviruses 6:999 (1990).

39. Robinson W.E., Montefiori D.C., and Mitchell W.M. Antibody dependent enhancement of human immunodeficiency virus type 1 infection. Lancet 1:790 (1988).

40. Robinson, W.E., Jr., Kawamura, T., Lake, D., Masuho, Y., Mitchell, W.M. and Hersh, E.M. Antibodies to the primary immunodominant domain of human immunodeficiency virus type 1 (HIV-1) glycoprotein gp41 enhance HIV-1 infection in vitro. J. Virol. 64(11):5301 (1990).

41. Robinson, W.E., Jr., Montefiori, D.C. and Mitchell W.M. Complement-mediated antibody-dependent enhancement of HIV-1 infection requires CD4 and complement receptors. Virology 175:600 (1990).

42. Halstead S.B., and O'Rourke E.J. Dengue viruses and mononuclear phagocytes. I. Infection enhancement by non-neutralizing antibody. J. Exp. Med. 146:210 (1977).

43. McKeating J.A., Griffiths P.D., and Weiss R.A. HIV susceptibility conferred to human fibroblasts by cytomegalovirus-induced Fc receptor. Nature 343:659 (1990).

44. Homsy J., Meyer M., and Levy J.A. Serum enhancement of human immunodeficiency virus (HIV) correlates with disease in HIV-infected individuals. J. Virol. 64:1437 (1990).

45. Hoxie J.A., Haggarty B.S., Rackowski J.L., Pilsbury N., and Levy J.A. Persistent noncytopathic infection of human lymphocytes with AIDS-associated retrovirus (ARV). Science 229:1400 (1985).

46. Folks T., Powell D.M., Lightfoote M.M., et al. Induction of HTLV-III/LAV from a non virus-producing T cell line: Implications for latency. Science 231:600 (1986).

47. Luciw P.A., Cheng-Mayer C., and Levy J.A. Mutational analysis of the human immunodeficiency virus (HIV): The orf-B region down-regulates virus replication. Proc. Natl. Acad. Sci. 84:1434 (1987).

48. Cheng-Mayer C, Ianello P., Shaw K, Luciw P., and Levy J.A. Differential effects of *nef* on HIV replication: Implications for viral pathogenesis in the host. Science 246:1629 (1989).

49. Kim S., Ikeuchi K., Byrn R., Groopman J., and Baltimore D. Lack of a negative influence on viral growth by the *nef* gene of human immunodeficiency virus type 1. Proc. Natl. Acad.Sci. 86:9549 (1989).

50. Hammes S.R., Dixon E.P., Maum M.H. Cullen B.R., and Greene W.C. *Nef* protein of human immunodeficiency virus type 1: Evidence against its role as a transcriptional inhibitor. Proc. Natl. Acad Sci. 86:9549 (1989).

51. Guy B., Kieny M.P., Riviere Y., Le Peuch C., Dott K., Girard M., Montagnier L. and Lecoco J.P. HIV F/3' orf encodes a phosphorylated GTP-binding protein resembling an oncogene product. Nature 330:266 (1987).

52. Poulin L. and Levy, J.A.*In vitro* expression of a functional HIV-**nef** gene product.V International Congress on AIDS,Montreal,1989, p 58.

MOLECULAR ANALYSIS OF RECEPTOR BINDING AND VIRAL TROPISM

A. Cordonnier and L. Montagnier

Unité d'Oncologie Virale (CNRS UA 1157)

Institut Pasteur, 75724 Paris Cedex 15, France

INTRODUCTION

Human Immunodeficiency Virus (HIV), the etiologic agent of AIDS (Acquired Immunodeficiency Syndrome) displays a selective tropism for CD4-positive lymphocytes and for CD4-positive cells of the mononuclear-phagocyte lineage. Infection of target cells is initiated by the specific interaction of HIV envelope protein gp120 and the CD4 molecule on the cell surface[1,2,3]. Despite extensive genetic variation of gp120 among different HIV isolates, each uses the CD4 molecule as a cellular receptor. Furthermore, both the HIV-1 and HIV-2 interact with the same epitopes on the CD4 molecule[4,5] while the extracellular envelope proteins of these two viruses display little antigenic cross-reactivity and share only 40% overall amino acid identity[6]. This suggests that conserved regions of gp120 are involved in binding to the CD4 molecule.

As a first step towards a functional analysis of the HIV envelope glycoprotein, we have constructed mutants by deleting conserved regions from the HIV-1 env gene[7]. The integrity of four domains is indispensable for binding to CD4 in vitro. In addition, some deletions interfere with maturation of the glycoprotein, but we found no correlation between altered glycosylation of the different mutant proteins and their capacity to bind to CD4. The results of these experiments are summarized in figure 1.

In a second step, we performed a detailed analysis of one of the regions that is important for binding to CD4. For this study, we choose region 5 (figure 1) because deletion of this region affect CD4 binding without interfering with maturation of the envelope protein. Furthermore, it has been shown by others that insertional mutagenesis at this site abrogates CD4 binding, and the epitopes for two antibodies that inhibit CD4-gp120 interaction map to this region[8]. We tested the effect of point mutations on CD4 binding and on viral infection of different cell lines[9]. Substitution of a single amino-acid can abrogate CD4 binding and viruses containing this mutation are non infectious. In contrast, other amino acid changes in the same region do not affect CD4 binding but alter viral tropism (table 1).

## EFFECTS OF DELETIONS IN HYPERCONSERVED REGIONS OF THE EXTRACELLULAR PROTEIN OF HIV-1

A comparison of the amino acid sequence of gp120 from the HIV-1 BRU isolate[10] with the gp120 of HIV-2 ROD[6] allowed us to define 6 regions that are conserved and homologous between HIV-1 and HIV-2 (regions boxed in figure 1) . To determine whether these regions are important for the virus-receptor interaction, we deleted nucleotides coding for the small stretches of amino acids (underlined in figure 1) in each of the conserved regions, without removing any cysteine residues. The mutated HIV-1 env genes were expressed in recombinant vaccinia viruses.

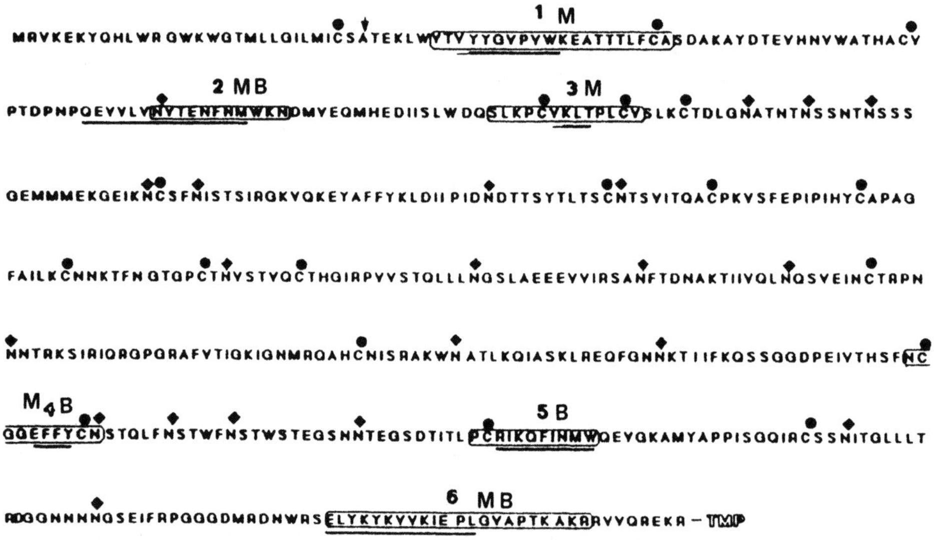


Figure 1
Amino acid sequence of the extracellular glycoprotein of HIV-1 (isolate BRU). The one letter amino acid abbreviation is used. The cysteine residues are marked by black dots and the potential site of glycosylation by open diamonds. Arrow indicate the position of the cleavage site between signal peptide and extracellular protein. TMP stands for the transmembrane protein. The 6 hyperconserved conserved between HIV-1 (BRU isolate) and HIV-2 (ROD isolate) are boxed. The amino acids deleted by site-directed mutagenesis are underlined.

The different deletions made in the gp120 env gene did not interfere with synthesis and stability of the glycoprotein, but some of them (indicated by a M in figure 1) altered glycoprotein processing and transport to the surface of infected cells. Subtle changes in the structure of the envelope protein affected its glycosylation even when a deletion did not alter any of the potential N-linked glycosylation site. Additional investigations will be aimed at a more precise definition of the relation between the pattern of glycosylation of each mutant protein and the step of maturation in which they it is blocked.

Binding to the CD4 molecule

To study receptor binding, we used a test based on the ability of a monoclonal antibody against CD4 (OKT4) to immunoprecipitate complexes of CD4 and HIV envelope proteins. After infection with recombinant vaccinia viruses producing non-deleted and deleted forms of the gp120, radio-labelled cell extracts were incubated with soluble CD4. The formation of gp120-CD4 complexes was assessed by immunoprecipitation with the OKT4 antibody. The specificity of this reaction was demonstrated using MAb OKT4A, which recognizes a CD4 epitope near the site of the gp120-CD4 interaction. This antibody failed to precipitate the complexes.

The CD4 binding assays show that deletions in at least four of the hyperconserved regions within gp120 (indicated by a B in figure 1) drastically reduce binding of the envelope protein to CD4. One of these regions is located within the first 100 amino acids of gp120 (region 2), while the others are near the C-terminus (regions 4, 5 and 6). Although these regions are distantly located with respect to each other in the primary sequence, it is likely that some are closely located in the three-dimensional structure of gp120 and contribute to a discontinuous binding site. Alternatively, since some deletions altered glycosylation of the envelope protein and its transport to the cell surface, it is possible that they also affect the tertiary folding of the protein and subsequently negate CD4 binding. However the fact that two differents monoclonal antibodies (Mabs 110-4 and 41-1; Genetic Systems) and a human polyclonal serum are able to precipitate all the mutant proteins suggests that they retain some native structure. It should be noted that the two mutants which were still able to bind CD4 had distinctly altered patterns of processing: Mutant 1 was not well expressed at the cell surface and mutant 3 reached the cell surface but gave rise to the smallest glycoprotein when produced with the recombinant vaccinia virus. The carbohydrate modification of these two mutants does not result in a lower affinity for the CD4 receptor. This result is consistent with other studies indicating that partially or extensively deglycosylated material is still able to bind to the receptor[11,12,13]. In contrast to these findings, use of glycosidase inhibitors in infected cultured cells demonstrated that envelope N-glycosylation is critical for the biological properties of HIV-1 at the levels of viral infectivity and cytopathogenicity[14,15,16]. Thus, it cannot be excluded that carbohydrates play a role in steps subsequent to viral binding such as fusion or that limited reduction in the affinity of gp120 to CD4 impairs the infectivity of viral particles. Nevertheless,a major and direct role of glycan moieties in gp120-CD4 interaction can be excluded.

The four regions of gp120 that we have shown to be important for CD4 binding are close to, but different from, those delineated by others using mutational analysis throughout gp120[17,18]. Epitope mapping of monoclonal antibodies that either block or don't block the CD4-gp120 interaction[8,19] constitute a

| | | | | | | | | | | | |
|---|---|---|---|---|---|---|---|---|---|---|---|
| **HIV-1** | | | | | | | | | | | |
| **BRU** | **C** | **R** | **I** | **K** | **Q** | **F** | **I** | **N** | **M** | **W** | **Q** |
| **HXB2,SF2 MAL,WMJ2 Z6,JY1** | — | — | — | — | — | **I** | — | — | — | — | — |
| **MN** | — | **K** | — | — | — | **I** | — | — | — | — | — |
| **SC** | — | — | — | — | **E** | **I** | — | — | — | — | — |
| **CDC4** | — | — | — | — | — | **I** | — | — | **R** | — | — |
| **RF** | — | — | — | — | — | **I** | **V** | — | — | — | — |
| **ELI** | — | — | — | — | — | **I** | — | **K** | — | **V** | **A** |
| **Z3** | — | — | — | — | — | **V** | **V** | **R** | **T** | — | — |
| **Z3Z1** | — | — | — | — | — | **I** | **V** | — | — | — | — |
| **NY5** | — | — | — | — | — | **I** | — | — | **S** | **G** | **R** |
| **HIV-2** | | | | | | | | | | | |
| **ROD** | — | **H** | — | — | — | **I** | — | — | **T** | — | **H** |
| **NIHZ** | — | — | — | **R** | — | **I** | — | — | **T** | — | **H** |
| **SIV** | | | | | | | | | | | |
| **SIVmac** | — | **H** | — | **R** | — | **I** | — | — | **T** | — | **H** |
| **SIVagm** | **Y** | **H** | — | **R** | **S** | **V** | — | — | **D** | **S** | **Y** |

Figure 2
Predicted amino acid sequence of hyperconserved region 5 of the extracellular glycoprotein of sequenced clones of HIV-1, HIV-2 and SIV.

complementary approach of interest. Taken together, different experimental approaches suggest that N-terminal and C-terminal portions of gp120 behave as structural components of the glycoprotein and are necessary to allow other parts of the molecule to form the tertiary structure of the binding site. In fact, Dowbenko et al.[8] have shown that monoclonal antibodies that block CD4-gp120 interaction recognize epitopes from residue 414 to 429 and residue 427 to 438 according to our nomenclature. These epitopes map to the same place as our region 5 (424-432). We therefore chose to analyse further this region that is highly conserved among isolates of HIV-1, HIV-2 and SIV (figure 2).

## EFFECTS OF POINT MUTATIONS WITHIN REGION 5

In order to map the residues of HIV-1 gp120 critical for CD4 binding and to analyze the effects of binding on viral infectivity, we introduced two short deletions and 13 amino acid substitutions within region 5. Mutated envelope genes were expressed by transfection with a vector in which the coding sequence for the transmembrane glycoprotein, gp41, was eliminated so that gp120 would be secreted into the tissue culture medium. This vector contains the HIV-2 long terminal repeat 5' to the env gene and is co-transfected with the HIV-1 tat gene. With this vector, envelope protein expression is independent of the HIV rev gene[20]. The CD4 binding activity of the mutant gp120 molecules was assayed by immunoprecipitation as described above. In order to determine whether the binding ability of soluble proteins in our assay was correlated with infectivity of viral particles, we introduced some of the mutations into an infectious HIV-1 proviral clone[21]. The results of CD4 binding in vitro by mutant gp120 molecules and infectivity of virions bearing these mutations are summarized in table 1.

We found that 6 of 15 mutant proteins bound to CD4 in a manner equivalent to native gp120. This indicates that at least some of the conserved amino acids of this region are not critical for receptor binding. However, both deletions of amino acids in this region and substitutions of the tryptophane at position 432 to either valine, glycine, serine or arginine eliminated CD4 binding. We are unable at this point to determine whether amino acid changes at tryptophan-432 affect the binding site directly, or whether they diminish binding by altering the conformation of another region of the molecule. It may be significant that even conservative amino acid substitutions of the tryptophan-432 such as phenylalanine or tyrosine decreased the CD4-binding capacity of gp120 by 80%. Although these latter mutants bind to CD4 weakly, viruses carrying such mutations were not infectious. Some of these substitutions occur in wild type isolates (figure 2) but not in infectious clones. Therefore, we do not yet know if the tryptophan at 432 can be compensated by other amino acids changes.

Interestingly, other amino acid changes in the same region had no measurable effect on binding but restricted viral

TABLE 1
Amino acid sequence, CD4 binding activity and viral infectivity of mutants in hyperconserved region 5 . Dashes indicate that an amino acid is identical to the BRU prototype shown at top. Delta 6 and delta 9 are deletions prepared by site-directed mutagenesis of the amino acids between the brackets. Mutations are named according to the wild-type amino acid and its position followed by the amino acid to which it has been changed.

| | | | CD4 Binding | Viral Infectivity in SUP-T1 | U937 |
|---|---|---|---|---|---|
| Wild-type : | HIV1-BRU | C R I K Q F I N M W Q E | ++++ | + | + |
| Mutants : | | | | | |
| Deletion 9 | | –[– – – – – – – – –]– – | – | – | – |
| Deletion 6 | | – –[– – – – – –]– – – – | – | | |
| Ile425-Thr | ( I425-T ) | – – T – – – – – – – – – | ++++ | + | – |
| Ile425-Lys | ( I425-K ) | – – K – – – – – – – – – | ++ | + | – |
| Lys426-Thr | ( K426-T ) | – – – T – – – – – – – – | ++++ | + | – |
| Lys426-Arg | ( K426-R ) | – – – R – – – – – – – – | ++++ | + | + |
| Gln427-Thr | ( Q427-T ) | – – – – T – – – – – – – | ++++ | + | – |
| Asn430-Tyr | ( N430-Y ) | – – – – – – – Y – – – – | ++++ | | |
| Asn430-His | ( N430-H ) | – – – – – – – H – – – – | ++++ | | |
| Trp432-Tyr | ( W432-Y ) | – – – – – – – – – Y – – | ++ | | |
| Trp432-Phe | ( W432-F ) | – – – – – – – – – F – – | ++ | – | – |
| Trp432-Ser | ( W432-S ) | – – – – – – – – – S – – | – | – | – |
| Trp432-Gly | ( W432-G ) | – – – – – – – – – G – – | – | | |
| Trp432-Val | ( W432-V ) | – – – – – – – – – V – – | – | | |
| Trp432-Arg | ( W432-R ) | – – – – – – – – – R – – | – | | |

tropism: virions containing mutations at position 425, 426 or 427 lost the ability to initiate a productive infection in cells of the monocyte lineage (U937 cells) while they were still able to infect T-lymphocytic cell lines (CEM, SUP-T1) and activated human peripheral blood lymphocytes. To ascertain that the binding to CD4 was not affected in these mutants, we performed a binding competition assay with excess of non-labeled purified gp120: the Ile425-Thr glycoprotein and wild-type gp120 showed a similar pattern while Trp432-Phe was competed out at lower gp120 concentration. Furthermore, Ile425-Thr gp120 showed the same pattern of binding to SUP-T1 cells and U937 cells as did wild-type gp120 (figure 3) indicating that mutated gp120 could bind not only soluble CD4 but also CD4 molecules anchored in membranes of SUP-T1 and U937 cells.

In conclusion, two major points can be made:

- First, some mutations in amino acids 425, 426 and 427 do not alter CD4 binding but affect another function required for viral entry. Indeed, transfection of U937 cells with proviral DNA containing the Ile425-Thr mutation leads to the transient production of viral particules, indicating that there is no restriction of replication or expression of mutated proviruses in these cells. Several other properties of gp120 such as association with the transmembrane glycoprotein,gp41, activation of the fusogenic domain of gp41 leading to the fusion of viral

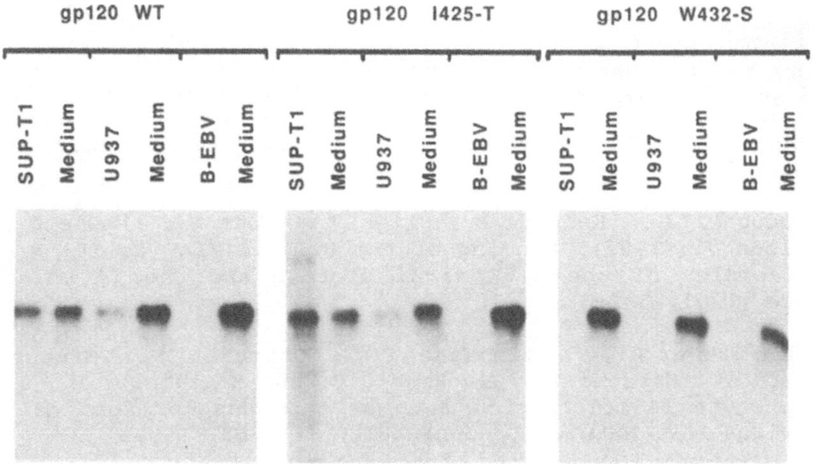


Figure 3
Binding of radio-labeled gp120 to CD4-positive cells. Supernatants from transfected COS-1 cells were radio-labeled and incubated with $10^7$ cells (SUP-T1, U937 or B-EBV transformed cells) for 2 hr at 4°C. The cells were washed 3 times in phosphate-buffered saline, lysed, and used for radioimmunoprecipitation with MAb 110-4 (Genetic Systems). Medium: unbound gp120 were also immunoprecipitated.

and cellular membranes, interaction with other molecules than CD4 on the cell surface and penetration of the plasma membrane are necessary to initiate a productive infection, after initial binding to the receptor. A similar functional defect was also found by others[22] that map to another region of the envelope protein. Molecular analysis of phenotypic revertants of the Ile425-Thr mutation will provide information about functional interactions of various regions of the envelope protein.

- Second, the effect of mutations at amino acid residues 425, 426 and 427 was drastic in the monocytic cell line U937 and not in the lymphocytic cell line SUP-T1, suggesting that the requirements for viral entry after initial binding are different in different cell lines. Although U937 cells may not be representative of macrophages in vivo, it is nonetheless tempting to correlate the restriction of tropism we obtained in this cell line with observations that natural variation of HIV affects the potential of the virus to grow in peripheral blood lymphocytes or macrophages. A differential tropism of HIV variants has been demonstrated depending on the tissue from which the virus was isolated[23,24]. In our situation, we can attribuate the restriction of tropism to a single amino acid change in gp120.

REFERENCES

1. Dalgleish, A., Beverly P., Clapham P., Crawford D., Greaves M., and Weiss R.. (1984), The CD4 (T4) antigen is an essential component of the receptor for the AIDS retrovirus. Nature (London) 312:763-766.

2. McDougal, J., Kennedy M., Sligh J., Cort S., Mawle A., and Nicholson J. (1986), Binding of the HTLV III/LAV to T4+ T cells by a complex of the 110K viral protein and the T4 molecule. Science , 231, 382-385.

3. Klatzmann, D., Champagne E., Chamaret S., Gruest J., Guetard D., Hercend T., Gluckman J.C. and Montagnier L. (1984), T-lymphocyte T4 molecule behaves as the receptor for the human retrovirus LAV. Nature (London), 312, 767-768.

4. Sattentau, Q. J., Dalgleish A. G. , Weiss R. A. and Beverly P. C. L. (1986), Epitopes of the CD4 antigen and HIV infection. Science, 234, 1120-1122.

5. Sattentau, Q. J. and Weiss R. A. (1988), The CD4 antigen: Physiological ligand and HIV receptor, Cell, 52, 631-633.

6. Guyader M., Emerman M., Sonigo P., Clavel F., Montagnier L. and Alizon M. (1987), Genome organisation and transactivation of the human immunodeficiency virus type 2. Nature (London), 326, 662-669.

7. Cordonnier A., Riviere Y., Montagnier L. and Emerman M. (1989), The effects of mutations in hyperconserved regions of the extracellular glycoprotein of HIV-1 on receptor binding. J.Virol., 63, 4464-4468.

8. Dowbenko D., Nakamura G. , Fennie C., Shimasaki C., Riddle L., Harris R., Gregory T. and Lasky L. (1988), Epitope mapping of the Human Immunodeficiency virus type 1 gp120 with monoclonal antibodies. J. Virol., 62, 4703-4711.

9. Cordonnier A., Montagnier L. and Emerman M. (1989), Single amino acid changes in the human immunodeficiency virus envelope affect viral tropism and receptor binding. Nature, 340, 571-574.

10. Wain-Hobson, S., Sonigo P. , Danos O., Cole S., and Alizon M. (1985), Nucleotide sequence of the AIDS virus, LAV. Cell , 40, 9-17.

11. Matthews, T.J., Weinhold K. J., Lyerly H. K., Langlois A. J., Wigzell H., and Bolognesi D. P. (1987). Proc. Natl. Acad. Sci. USA, 84, 5424-5428.

12. Fennie, C. and Lasky, L. A. (1989). Model for intracellular folding of the human immunodeficiency virus type 1 gp120. J.Virol., 63, 639-646.

13. Fenouillet E., Clerget-Raslain B., Gluckman J. C., Guetard D., Montagnier L. and Bahraoui M. (1989), Role of N-linked glycans in the interaction between the envelope glycoprotein of HIV and its CD4 cellular receptor. Structural enzymatic analysis. *J. Exp. Med.*,

14. Dewar, R.L., Vasudevachari, M. B., Natarajan V. and Salzman N. P. (1989), Biosynthesis and processing of human immunodeficiency virus type 1 envelope glycoproteins: Effects of monensin on glycosylation and transport. J. Virol., 63, 2452-2456.

15. Montefiori D. C., Robinson W. E. and Mitchell W. M. (1988), Role of protein N-glycosylation in pathogenesis of human immunodeficiency virus type 1. Proc. Natl. Acad. Sci. USA, 85, 9248-9252.

16. Pal R., Gallo R. C. and Sarngadharan M. G. (1988), Processing of the structural proteins of human immunodeficiency virus type 1 in presence of monensin and cerulenin. Proc. Natl. Acad. Sci. USA, 85, 9283-9286.

17. Kowalski, M., J. Potz, L. Basiripour, T. Dorfman, W. C. Goh, E. Terwilliger, A. Dayton, C. Rosen, W. Haseltine, and J. Sodroski (1987). Functional regions of the envelope glycoprotein of the human immunodeficiency virus type 1. Science 237:1351-1355.

18. Lasky, L. A., G. M. Nakamura, D. H. Smith, C. Fennie, C. Shimasaki, E. Patzer, O. Berman, T. Gregory, and D. J. Capon.. 1987. Delineation of a region of the human immunodeficiency virus type 1 gp120 glycoprotein critical for interaction with CD4 receptor. Cell 50:975-985.

19. Linsley, P. S., Ledbetter J. A., Kinney-Thomas E., and Hu S. L. (1988), Effects of anti-gp120 monoclonal antibodies on CD4 receptor binding by the env protein of human immunodeficiency virus type 1. J. Virol., 62, 3695-3702.

20. Emerman, M., Vazeux, R. & Peden, K. (1989). The rev gene product of the human immunodeficiency virus affects envelope-specific RNA localization. Cell, 57, 1155-1165.

21. Adachi, A., H.E. Gendelmann, S. Koenig, T. Folks, R. Willey, A. Rabson, and M.A. Martin. 1986. Production of acquired immunodeficiency syndrome-associated retrovirus in human and non-human cells transfected with an infectious molecular clone. J. Virol. 59: 284-291.

22. Willey, R. L., D. H. Smith, L. A. Lasky, T. S. Theodore, P.L. Earl, B. Moss, D. J. Capon, and M. A. Martin. 1988. In vitro mutagenesis identifies a region within the envelope gene of the human immunodeficiency virus that is critical for infectivity. J. Virol. 62:139-147.

23. Gartner, S., Markovits, P., Markovitz, D. M., Kaplan M. H., Gallo, R. C., Popovic M. (1986). The role of mononuclear phagocytes in HTLV-III/LAV infection. Science 233, 215-219.

24. Koyanagi, Y., Miles S., Mitsuyasu R. T., Merrill J. E., Vinters H.V., Chen I. S.(1988).Science 236, 819-822.

ACKNOWLEDGEMENTS
This work was done with the collaboration of M.Emerman and Y. Rivière.

# CELLS INFECTED BY HUMAN IMMUNODEFICIENCY VIRUS IN VIVO

Catherine Reynolds-Kohler, Clayton Wiley and Jay A. Nelson

Department of Immunology
Research Institute of Scripps Clinic
La Jolla, CA
Department of Pathology
University of California San Diego
La Jolla, CA

## INTRODUCTION

The identification of human immunodeficiency virus (HIV) as the etiologic agent of AIDS has permitted the characterization of the virus by *in vitro* culture analysis. The ability to make HIV nucleic acid and antibody probes has also allowed us to determine the involvement of the virus in disease mechanisms *in vivo*. Classically, HIV is known to have a preferential tropism for cells that express the CD4 receptor[1]. These cells include T helper cells and many cells of the monocyte-macrophage lineage[1-4]. In addition to T cells and monocyte/macrophages, a variety of other cell types have been shown to support the growth of HIV including B-cells, Langerhans cells, follicular dendritic cells, glial cells and cells of the colon[5-9]. In order to understand the mechanism fo HIV infection it is important to identify the cell types naturally infected with the virus. The following chapter describes cell types that are infected in three distinct tissues. By utilizing *in situ* hybridization and immunocytochemistry techniques we have analyzed brain, bowel and placental tissue for the presence of HIV nucleic acid and antigen. In this chapter we applied these techniques to identify key cells naturally infected by HIV and to determine a possible mechanism for the transmission of the virus.

## BRAIN

### Central nervous system disease associated with HIV infection

A multitude of neurologic disorders have been noted in patients with AIDS. In fact, between 20 and 30% of adults with AIDS or AIDS Related Complex (ARC) have neurologic symptoms attributable to HIV infection of CNS[10-13]. These neurologic symptoms include decreased memory, termed AIDS dementia complex (ADC)[13], a decrease in the ability to concentrate and motor and behaviorial disturbances. Nevertheless, it is still difficult to link the clinical encephalopathy with specific pathologic processes. One-third of AIDS autopsies have a viral encephalopathy diagnosed as "subacute encephalitis"[10,11,14].

The most noted neuropathologic finding in the brain tissue from patients with ADC is a diffuse gliosis in the basal ganglia and dep white matter. Other common findings include microglial nodules, multinucleated giant cell foci, myelin pallor, and collections of perivascular chronic inflammatory cells[15-17].

The direct cause of the central nervous system (CNS) histopathology and the dementia in AIDS patients is still under investigation. There may be several potential mechanisms acting either independently or synergistically to produce the CNS destruction commonly found in patients with AIDS. Damage to the CNS could occur by the direct infection of HIV in the brain tissue[18]. On the other hand, a variety of pathogenic viruses could also be contributing to the histopathology[19]. Cofactors of CNS pathology could also include bacterial, fungal, and protozoan opportunistic pathogens. Alternatively, CNS injury could result from indirect effects of systemic or local viral infection caused by viral or immune factors.

### Cellular localization of HIV

As early as 1985, it was observed that there was a significant distribution of HIV infected tissues in AIDS patients[18]. One of the surprising results from these experiments was the localization of HIV in the brain. HIV could easily be isolated from the cerebral spinal fluid (CSF) and neural tissue of AIDS patients[12,20]. These observations suggested that the CNS was a potential reservoir for the AIDS virus and might be able to explain some of the neurologic symptoms associated with the disease. Histological identification of the virus by our group and others, utilizing immunocytochemistry and *in situ* hybridization, definitively localized HIV primarily within cells of monocyte/macrophage lineage[21-27]. However, other cell types less frequently infected have also been observed in these tissues.

Between 1982-1987 our group examined brain tissue from 93 autopsies. A summary of the HIV infected cell types and the relative frequency of HIV antigen or nucleic acid found in each of the 37 cases is supplied in Table 1. Macrophages and multinucleated giant cells were the cell type most commonly observed infected with the virus (Figs.1a and b). In approximately half of the cases, endothelial cells were also found to be infected (Fig.1c). Only one case, in which there was severe CNS destruction, could we verify virus in the CNS parenchymal cells (Fig.1d).

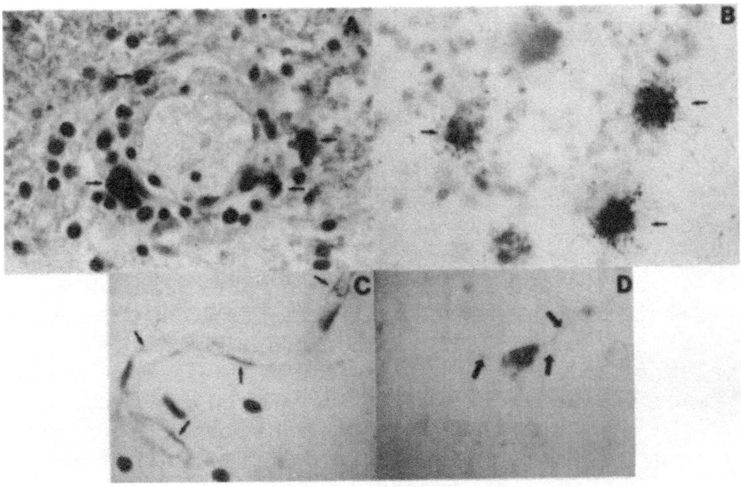


Fig. 1. Paraffin embedded sections of brain autopsies of AIDS patients. a) Section stained immunocytochemically with HIV antibody demonstrating perivascular macrophages containing HIV antigen. b) *In situ* hybridization with a $^{35}S$ labeled HIV probe demonstrates grains over multinucleated giant cells. c) Section stained immunocytochemically demonstrates HIV infection of endothelial cells. d) Section stained immunocytochemically demonstrates a rare cell infected by HIV with spider-like processes (arrow). While the lineage of such cells cannot be precisely determined, the morphology is consistent with glial cells.

Although neuronal cells and glial cells have been readily infected with HIV in tissue culture systems, *in vivo* results suggest that these cell types are not targets for infection by the virus[8,28]. These cells have been shown to possess a truncated CD4 mRNA, yet experiments using monoclonal antibodies to block the action of CD4 do not inhibit *in vitro* HIV infection[8,28,29]. These observations suggest that the CNS dysfunction observed in AIDS patients may not involve direct HIV infection of neurons and glial cells. However, the presence of HIV in CNS tissue may have indirect effects resulting in the damage typically observed in these individuals.

Table 1. Cellular Localization of HIV in the Central Nervous System of 37 Autopsied AIDS Cases*

| | Rare | Occasional | Frequent | Total |
|---|---|---|---|---|
| Macrophages | 7 | 12 | 17 | 36 |
| Endothelial cells | 1 | 12 | 8 | 21 |
| Cells of unknown lineage possibly oligodendrocytes | Not assessed | 13 | 1 | 14 |

Mechanisms of transmission

Cells of the monocyte/macrophage lineage are the most frequently HIV infected cell type in the brain. It is controversial whether these cells are microglia or represent monocytes/macrophages that have migrated into the CNS[30]. The microglia are resident brain macrophages which have a hematogeneous origin. In AIDS patients with CNS disease, the perivascular location of HIV infected cells suggests a hematogenous origin of the infection (Fig. 1a). The amount of monocyte/macrophages in the brain which contain viral antigen and nucleic acid is incommensurate to the amount of cells which contain virus in the peripheral blood[31]. Recently, we have found that high percentages of monocytes from HIV seropositive donors can be induced to replicate HIV by cell contact with activated T cells[32]. These observations suggest that a specific differentiation pathway is necessary in monocytes/macrophages to obtain activation of HIV. It has been demonstrated in other virus systems that monocyte/macrophages induced to differentiate can release quiescent virus[33]. Perhaps a similar sequence of events may occur during HIV infection of the peripheral blood. Circulating monocytes that are latently infected with HIV may transport the virus to the brain. Upon cell differentiation in the brain, the virus could then be expressed and be transmitted to other cells. This scenario may best be exemplified by JC virus infections of the brain. JC virus is frequently found in AIDS patients and predominantly affects the CNS[34-37]. Lytic infection of oligodendrocytes by JC virus causes a demyelinating disease called progressive multifocal leukoencephalopathy. Infiltration of brain tissue by macrophages is a characteristic of JC virus infection. Brain biopsies of PML patients demonstrated HIV-infected macrophages infiltrating regions of demyelination[36]. Upon death of these patients, examination of the brain tissue revealed increased numbers of HIV infected macrophages. A significant percentage of these macrophages fused into multinucleated giant cells. These findings suggest that JC virus not only directly infects oligodendrocytes to cause neuronal damage, but also contributes to the influx of macrophages into the brain which is then susceptible to HIV infection.

Another common viral pathogen found in the CNS of AIDS patients is cytomegalovirus (CMV). In our studies of 93 autopsies from AIDS patients, 60% of the patients with subacute encephalitis were infected by CMV[36]. CMV may act in a manner similar to other opportunistic infections by attracting HIV-infected monocytes into the CNS. In most cases, CMV is widely disseminated throughout the brain, a pattern of CMV infection similar to that seen in other immunocompromised hosts. However, in some AIDS patients CMV infection is more fulminant and the pattern of spread unique to AIDS. These cases of fulminant CMV growth are characterized by an increase of HIV infected macrophages surrounding the CMV infected areas. The high amounts of HIV seen in the presence of CMV suggest synergistic interactions between the two viruses. Importantly, coinfection of the same cell by HIV and HCMV has been detected in the brain tissue of AIDS patients. This observation implies another level of cooperativity. Infection by one virus might change host-cell permissiveness to infection by a second virus through viral transactivator proteins. It has recently been shown that the immediate early genes produced by CMV can transactivate HIV expression[37].

The most perplexing phenomenon involving the infection of HIV in the brain is the discrepancy between the level of CNS destruction and the amount of virus detected. This could simply be a problem of antigen and nucleic acid degradation due to the variability in preserving the tissue samples. Thus, the amount of virus we are detecting with our current methods may not be an accurate representation of the actual amount of virus present to cause the tissue destruction. On the other hand, it may be possible that there is not a one-to-one correlation between the amount of virus and the actual CNS damage. In this case, the nervous sytem damage may be amplified through an indirect mechanism. This indirect mechanism could include retroviral toxicity to the neuronal cells, blood brain barrier disruption due to endothelial cell infection, and CNS destruction mediated by the immune system itself[4].

Another important consideration is the differential expression of HIV CNS isolates in various cells. Different isolates of HIV have been shown to have variable tropisms for cells in the CNS[8]. Both genetic and phenotypic differences have been observed among HIV isolates derived from the CNS[38]. In this report isolates from patients with neurologic manifestations were found to be less cytocidal than those derived from other sources. Whether these differences reflect the ability of HIV to infect CNS tissue is unclear.

## BOWEL

### Disease associate with HIV infection

Individuals infected by HIV can suffer various gastrointestinal symptoms including chronic diarrhea. The cause of this problem was often a known infectious agents, but in several cases no pathogen could be identified. In Africa, this disorder has been associated with slim disease, a term that describes the great weight loss that accompanies diarrhea[39,40].

Other groups have found histological abnormalities in the small intestine and bowel of patients infected with HIV[41,42]. These abnormalities include low-grade small bowel atrophy and maturational defect in enterocytes. Brush border enzymes in many HIV infected patients also showed decreased activity. The decrease of an enzyme called B-glucosidase found in many AIDS patients could explain many of the maladsorption problems[41]. In the absence of detectable enteropathogens, crypt hyperplastic atrophy has been detected at all stages of HIV infection[42]. Apoptoses, or individual cell necrosis in crypt epithelium is a characteristic histologic abnormality in HIV infection[43]. This disorder is thought to occur via a mechanism very similar to graft vs host disease[42].

HIV was shown to infect human colorectal cell lines in 1987[39]. In these rectal epithelial cells infected by HIV, CD4 mRNA but not protein was detected in the cells. This group also demonstrated similar amounts of CD4 mRNA in normal colon mucosa, suggesting that epithelium in the digestive tract possesses a receptor for HIV. These results implied that HIV may have a tropism for another tissue type. Our results[44,45] and the results of others[46] show that normal bowel mucosa can be infected by HIV.

### Cellular localization of HIV

In our studies to determine whether bowel tissue contains infectious HIV, rectal mucosal biopsy specimens were taken from four HIV seropositive patients with a history of chronic watery diarrhea[44]. After all contaminating blood was thoroughly removed by extensive washing, infectious HIV was recovered from two out of the four biopsy specimens.

To identify cell types naturally infected by HIV, we examined bowel biopsy material by *in situ* hybridization and enzyme histocytochemistry. Five of the ten biopsies showed detectable nucleic acid in the rectal and duodenal tissues. The region of the bowel from AIDS patients which showed the highest number of HIV infected cells by *in situ* hybridization was the rectum. In most cases, HIV hybridization occurred at the base of the bowel crypts (Fig. 2), although in some cases (both rectal and duodenal tissue) HIV nucleic acids were detected in the lamina propria. Immunocytochemistry using antibody against HIV confirmed the *in situ* hybridization results.

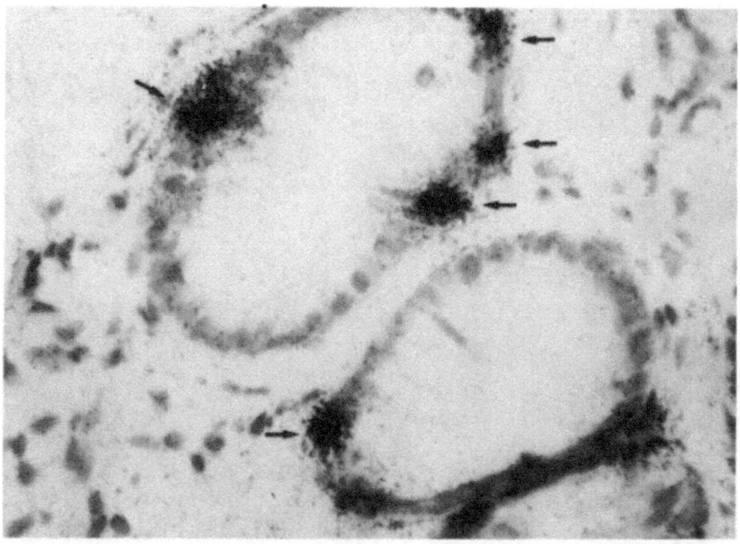

Fig. 2. Paraffin embedded section of bowel biopsies taken from AIDS patients with no known cause of the symptoms. *In situ* hybridization was performed with a $^{35}S$-labeled HIV probe. Silver grains indicating areas of hybridization were detected in the crypts of the bowel.

Several cell types are found in the bowel crypts including columnar epithelial and goblet cells, migratory intraepithelial mononuclear cells, and enterochromaffin cells. Columnar epithelial and goblet cells are found on the luminal surface of the crypts. Detection of viral nucleic acids, specifically in the base of the crypts, argues against either epithelial cells or goblet cells as the target for infection. Therefore, enterochromaffin cells seemed a potential candidate for HIV infection. These cells are derived from neural crest cells and regulate fluids in the bowel.

To determine whether HIV infects enterochromaffin cells, we utilized argentaffin staining for identification. Argentaffin stain is an enzyme histochemical method for detecting reducing substances in enterochromaffin cells. Initial experiments showed that cells in the crypt which were hybridizing to HIV nucleic acid were also staining with the argentaffin stain. Therefore, to establish that the cells staining with argentaffin were in fact the same as those hybridizing to HIV, we performed a double-labeling experiment. Cells were first treated with the argentaffin stain and then hybridized with an HIV specific probe. Not all enterochromaffin cells that stained hybridized with the probe; however, a small percentage of cells were double-labeled (Fig. 3). In addition to enterochromaffin cells, other crypt cells hybridized with the HIV probe. These cells included the columnar epithelial and the goblet cells.

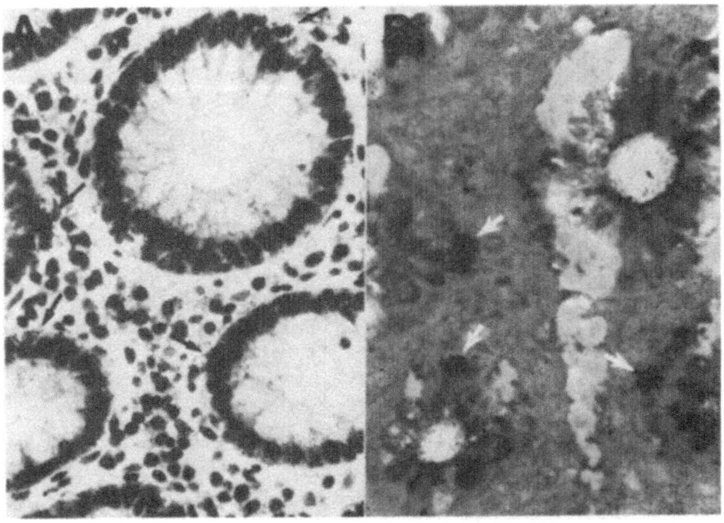

Fig. 3. The section of paraffin-embedded rectal mucosa tissue was first examined by *in situ* hybridization, and then stained by Schmorl's method to detect the presence of enterochromaffin cells. Two of the cells that stained as enterochromaffin cells also demonstrated the presence of HIV (silver grains). Goblet cells containing mucosa stained by mucicarmine.

Mechanisms of transmission

The observation that transformed bowel cells are infected by HIV strongly suggests that these cells possess a receptor, possibly CD4, that allows the infection to occur[9]. Consistent with this data, infectious HIV can be recovered from rectal mucosal biopsy specimens of AIDS patients. In light of this evidence, bowel cells should be regarded as potential sites for direct infection. HIV infection of bowel tissue is important when one considers the risk of transmission during anal/genital contact. Infectious virus has been recovered from seminal fluid and therefore may directly infect epithelial cells in the rectal mucosa. However, the detection of HIV infected cells in the lamina propria suggests that hematogenous spread may also occur.

HIV infection of the bowel may also have indirect effects contributing to the disease process. In the normal bowel the intestinal lamina propria has been shown to contain a higher percentage of CD4 positive cells[47,48] than in patients with AIDS[49-52]. Depletion of CD4 positive lymphocytes may cause impaired mucosal immune function leading to the growth of secondary infections in the gastrointestinal tract. In addition, recent studies have shown that the intestinal lamina propria contains specialized activated T cells[48]. These T cells may be involved in the maintenance of the normal mucosa in the bowel. Depletion of these cells due to HIV infection may cause indirect damage to the bowel. HIV infects a variety of cells in the bowel. Further studies will determine whether the gastrointestinal disease suffered by some AIDS patients directly results from HIV infection or is a secondary effect.

**PLACENTA**

Disease associated with HIV infection

HIV infection in infants is most frequently acquired by vertical transmission[53]. Thirty to fifty percent of infants born to mothers who are infected with HIV become infected[54,55]. In New York City, statistics demonstrate that HIV infects infants at a ratio of 1 in 200 births. Twenty percent of the infants infected with HIV die before 18 months of age.

It is still debatable whether HIV causes abnormal fetopathy. Some groups have shown a multitude of abnormalities including severe growth failure, microcephaly, ocular hypertelorism and severe facial disfiguration[56]. To further define the features of dysmorphology associated with the congenital infection by HIV, these groups set up a scoring system called the "fetal AIDS syndrome score". In one case, the diagnosis of intrauterine HIV infection was made in a routine examination on the basis of the dysmorphic facial features[57]. A group in New York City refute these findings and in their studies found no significant differences in the incidence of craniofacial dysmorphism[58]. There is still no consensus regarding this issue.

By the age of two, most congenitally infected infants demonstrate signs of immunodeficiency. These children have many illnesses including pneumonia, meningitis and bone and joint problems, although opportunistic infections are less common than in adults.

Cellular localization of HIV

Evidence of *in utero* HIV infection in second and third trimester fetal tissue has been shown by several groups[59-61]. Two of these groups were able to recover virus from the fetal tissue by co-cultivating the tissues with lymphocyte cell lines[60,61]. The other group was able to detect HIV antigens in thymic T cells early in the third trimester[59].

In order to understand and explain the mechanism of *in utero* HIV infection, our group analyzed placental tissue from abortuses of HIV seropositive mothers. Tissues were examined by both *in situ* hybridization and immunocytochemistry in order to identify the specific cell types involved in the transplacental passage of HIV. We identified HIV antigen and nucleic acid in tissue from three fetuses aborted at eight weeks. Maternal decidual leukocytes, trophoblast cells, villous mesenchymal cells (macrophages and endothelial cells) and embryonic blood cell precursors were all infected by HIV. In fetal tissues the trophoblast layer of the placenta demonstrated the highest frequency of staining and hybridization (Fig. 4a). The trophoblast which surrounds the chorionic villi is composed of the outer syncytiotrophoblast layer and the inner cytotrophoblast layer. The syncytiotrophoblast layer had a much higher frequency of staining and hybridization with HIV antibodies and probes as compared to the cytotrophoblast (Fig. 4a). The villous macrophages or Hofbauer cells were also frequently observed to contain HIV antigen and nucleic acid as well as occasional nucleated blood precursor cells (Fig. 4b).

Transmission of the virus

The most likely cell types to be involved in the transmission of the HIV across the placental barrier are the trophoblast derivatives, specifically the syncytiotrophoblast cells. These cells are the pivotal cells at the interface between the maternal/fetal circulation. It is still unknown whether or not these cells possess CD4. Immunolabeling studies of cultured trophoblast from first trimester fetuses have shown CD4 positive along the perimeter of chorionic villi[62]. We, however, have been unable to detect the presence of the antigen in placental sections by immunohistochemistry (unpublished observations). Phagocytosis has been shown to be a function of syncytiotrophoblast cells which could explain the mechanism of HIV passage from the mother to the fetus in the absence of CD4[63].

From this information, three possible mechanisms for HIV infection of fetal tissue exists. First, free virus released by the decidual leukocytes can bind or be phagocytosed by the syncytiotrophoblasts. Second, the trophoblasts could be infected by direct contact with HIV infected decidual leukocytes. This infection could occur by some receptor mediated mechanism or by phagocytosis. Finally, the virus could spread as infected maternal leukocytes invade the villous stroma.

Once the virus has infected the syncytiotrophoblast cells, the next question is how the virus can gain access from these cells to the fetus. The placental macrophages or the Hofbauer cells bear the CD4 receptor as well as other macrophage markers[64]. Direct infection of these cells by contact with HIV infected trophoblast cells could occur. This evidence is important in light of the fact that Hofbauer cells have been shown to be involved in trophoblast homeostasis[65]. The Hofbauer cells, once infected, may then infect villous capillary endothelium. Alternatively, the infected Hofbauer cells could diapedese into the capillary luminae and directly infect the blood cell precursors. It is still unknown whether these blood cell precursors possess the CD4 receptor and can be directly infected by the virus.

The above observations suggest that the passage of HIV proceeds from maternal decidual leukocytes to the trophoblast cells, specifically the syncytiotrophoblasts. HIV infected trophoblast cells can then infect Hofbauer cells, possibly via the CD4 receptor. The Hofbauer cells can then infect the villous endothelium or directly infect blood cell precursors. This model is based on HIV nucleic acid and antigen distribution in the placental tissue at eight weeks gestational age.

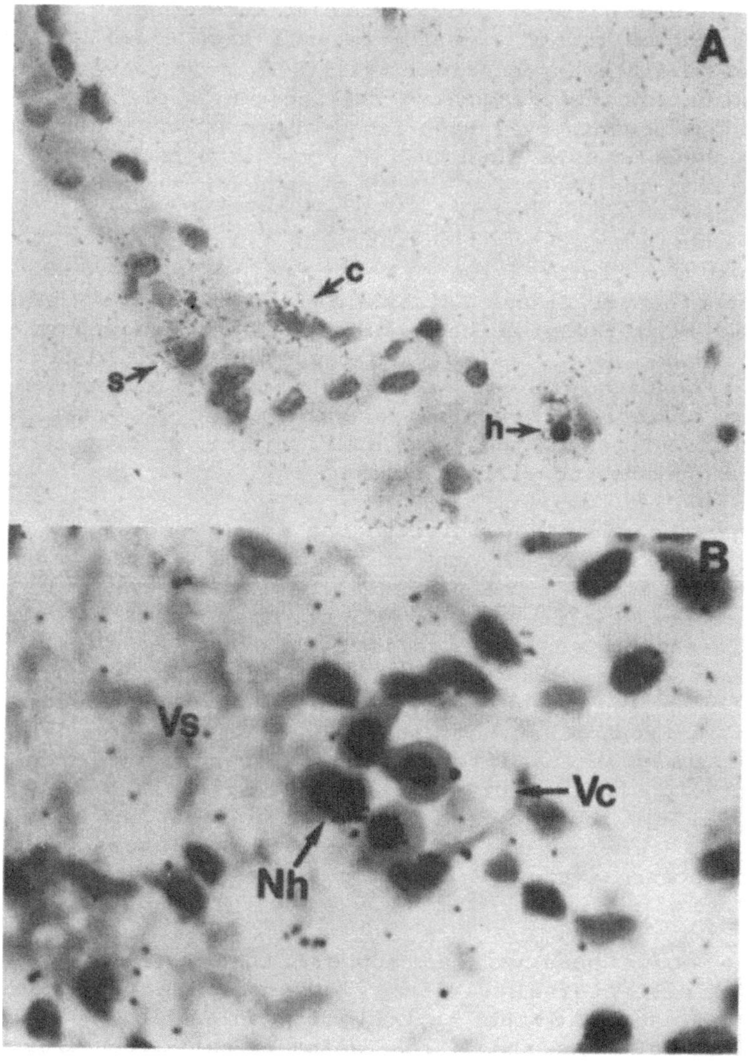


Fig. 4. Paraffin-embedded sections of tissue of fetuses aborted at 8 weeks from an HIV-seropositive mother hybridized with a $^{35}S$-labeled HIV probe. a) Note labeling of syncytiotrophoblast (s) and cytotrophoblast (c) as well as a Hofbauer cell (h). b) Note labeling of nucleated precursor cell (Nh), villous stroma (Vs) and a vessel within the villous (Vc) are also depicted.

## CONCLUSION

We have shown evidence, in three distinct tissues, that HIV has a diverse tropism. In brain, bowel and placenta a variety of cell types can be infected with HIV. Whether or not cells possess the CD4 receptor, HIV nucleic acid and antigen could still be detected. These results demonstrate the fact that there may be other mechanisms by which the virus can infect cells.

Other factors that are important in HIV infections include the presence of opportunistic pathogens as well as the effect of immune factors. Opportunistic viral pathogens have been shown to act synergistically with HIV to enhance or facilitate infection. In the absence of CD4, antibodies may facilitate the attachment of HIV to the target cells through Fc receptors.

Whether or not a specific cell type is infected with HIV is dependent on a number of variables. We have presented evidence showing different cell types infected by HIV, as well as some possible theories on the transmission of the virus in these tissue types. Further characterization of these cell types will aid in understanding the mechanisms of infection by HIV. This information is vital to aid in clinical intervention and preventive therapy.

## REFERENCES

1. D. Klatzmann, F. Barre-Sinoussi, M.T. Nugeyre, C.Dauguet, E. Vilmar, C. Griscelli, F. Brun-Vezinet, C. Rouzioux, J.C. Gluckman, J.-C. Chermann, and L. Montagnier, Selective trophism of lymphadenopathy associated virus (LAV) for helper-inducer lymphocytes. Science 225:59 (1984).
2. Q.J.Sattentan, A.G. Dalgleish, R.A. Weiss, and P.C.C. Beverly, Epitopes of the CD4 antigen and HIV infection. Science 234:1120 (1986).
3. S. Gartner, P. Markovitz, D.M. Markovitz, M.H. Kaplan, R.C.Gallo, and M. Popovic, The role of mononuclear phagocytes in HTLV-III/LAV infection. Science 233:215 (1986).
4. D.D.Ho, T.R.Rota, and M.S.Hirsch, Infection of monocyte/ macrophages by human T lymphotropic virus type III. J. Clin. Invest. 77:1712 (1986).
5. L. Montagnier, J. Gruest, S. Chamaret, D. Klatzman and J.C. Gluckman. Adaptation of lymphadenopathy associated virus (LAV) to replication in EBV-transformed B lymphoblastoid cell lines. Science 225: 63 (1984).
6. E. Tschachler, V. Groh, M.Popovic, et al. Epidermal Langerhans cells--a target for HTLV-III/LAV infection. J. Invest. Dermatol. 88:233 (1988)

7. K. Rappersberger, S. Gartner, P. Schenk, et al. Langerhans cells are an actual site of HIV-1 replication, Intervirology 29:185 (1988).
8. C. Cheng-Mayer, J.T. Rutka, M.L. Rosenblum, T. McHugh, D.P. Stites, J.A. and Levy, Human immunodeficiency virus can productively infect cultured human glial cells. Proc. Natl. Acad. Sci. 84:3526 (1987).
9. A. Adachi, S. Koenig, D. Daugherty, S.Gatton-Celli, A.S. Fauci, M.A. Martin, and H.E. Gendelman, Productive persistent infection of human colorectal cell lines with human immunodeficiency virus. J. Virol. 61:209 (1987).
10. W.D. Snider, D.M. Simpson, S. Nielson, J.W.M. Gold, C.E. Metrols, and J.B.Posner, Neurological complications of acquired immune deficiency syndrome: Analyses of 50 patients. Ann. Neurol. 14:403 (1983).
11. S.L. Nielson, C.K. Petito, C.D. Urmaeher, and J.B. Posner, Subacute encephalitis in acquired immune deficiency syndrome: A postmortem study. Am. J. Clin. Path. 82:678 (1984).
12. J.A. Levy, J.Shimabukuro, H. Hollander, J. Mills, and L. Kaminsky, Isolation of AIDS-associated retroviruses from cerebrospinal fluid and brain patients with neurologic symptoms. Lancet ii:586 (1985).
13. B.A. Navia, E-S.Cho, C.K. Petito, and R.W. Price, The AIDS dementia complex II. Neuropathology. Ann. Neurol. 19:525 (1986).
14. L.B. Moskowitz, G.T. Hensley, J.C. Chan, J. Gregorios, and F.K. Conley, The neuropathology of acquired immune deficiency syndrome. Arch. Path. Labs. Med. 108:867 (1984).
15. A.A. Harry, J. Segneti, and S. Leven, Central nervous system infections in patients with acquired immune deficiency syndrome (AIDS). Clin. Neuropharmacol. 8:201 (1984).
16. R.M. Levy, D.E. Bredsen, and M.L.Rosenblum, Neurological manifestations of the acquired immunodeficiency syndrome (AIDS) experience at UCSFD and review of the literature. J. Neurosurg. 62:475 (1985).
17. K.H. Anders, W.F. Giverra, U. Tomiyasu, M.A. Verity, and H.V. Vinters, The neuropathology of AIDS. Am. J. Path. 124:537 (1986).
18. G.M. Sjaw, M.D. Harper, B.H. Hahn, L.G. Epstein, D.C. Gajdusek, R.W. Price, B.A. Navia, C.K. Petito, C.J. O'Hara, E-S. Cho, J.M. Oleske, F. Wong-Staal, and R.C.Gallo, HTLV-III infection in brains of children and adults with AIDS encephalopathy. Science 227:177 (1985).
19. C.A. Wiley, and J.A. Nelson, Human immunodeficiency virus: Infection of the nervous system, in "Current Topics in Microbiology and Immunology Retrovirus infections of the nervous system". M.B.A. Oldstone and H. Koprowski, eds., Springer Verlag, Heidelberg. (1990).

20. D. Ho, T.R. Rota, R.T. Schooley, J.C. Kaplan, J.D. Allan, J.E. Groopman, L. Resnick, D. Felsenstein, C.A. Andrews, and M.S. Hirsch, Isolation of HTLV III from cerebrospinal fluid and neural tissue of patients with neurologic syndromes related to the acquired immunodeficiency syndrome. N. Engl. J. Med. 313:1493 (1985).
21. C.A. Wiley, R.D. Schrier, J.A. Nelson, P.W. Lampert, and M.B.A. Oldstone, Cellular localization of human immunodeficiency virus infection within the brains of acquired immune deficiency patients. Proc. Natl. Acad. 83:7089 (1986).
22. M.H. Stoler, T.A. Eskin, S. Benn, R.C. Angerer, and L.M. Angerer, Human T-cell lymphotrophic virus type III infection of the central nervous system. JAMA 256:2360 (1986)
23. D.H. Gabuzda, D.D. Ho, S.M. de la Monte, M.S. Hirsch, T.R. Rota, and R.A. Sobel, Immunohistochemical identification of HTLV III antigen in brains of patients with AIDS. Ann. Neurol. 20:289 (1986).
24. S. Koenig, H.E. Gendelman, J.M. Orenstein, M.C. Dal Canto, G.H. Pezeshkpour, M. Yungbluth, F. Janotta, A. Aksanut, M.A. Martin, and A.S. Fauci, Detection of AIDS virus in macrophages in brain tissue from AIDS patients with encephalopathy. Science 233:1089 (1986).
25. J.M. Ward, T.J. O'Leary, G.B. Baskin, R. Benveniste, C.A. Harris, P.L. Nara, and R. Rhodes, R., Immunohistochemical localization of human and simian immunodeficiency viral antigens in fixed tissue sections. Am. J. Pathol. 127:199 (1987).
26. R. Vazeux, N. Brousse, A. Janry, D. Henin, C. Marche, C. Vedrenne, J. Mikol, M. Wolff, C. Michon, W. Rozenbaum, J-F. Bureau, L. Montagnier, and M. Brahic, AIDS subacute encephalitis: Identification of HIV infected cells. Am. J. Pathol. 126:403 (1987).
27. S.W. Rostael, S.M. Sumi, C.M. Shaw, K. Olson, and J.K. McDougall, Human immunodeficiency virus (HIV) infection in brains with AIDS-related leukoencephalopathy. AIDS Res. Hum. Retro. 3:4 (1987).
28. J. Weber, P. Clapham, J. McKeating, M. Stratton, E. Robey, and R. Weiss, Infection of brain cells by diverse immunodeficiency virus isolates: Role of CD4 as receptor. J. Gen. Virol. 70:2653 (1989).
29. I. Funke, A. Hahn, E.P. Rieber, E. Weiss, and G. Reithmuller, The cellular receptor (CD4) of the human immunodeficiency virus is expressed on neurons and glial cells in human brain. J. Exp. Med. 165:1230 (1987).
30. E-A. Ling, in "Advances in Cellular Neurobiology", S. Federoff and L. Hertz, eds., Academic Press, New York.
31. M.E. Harper, L.M. Marselle, R.C. Gallo, and F. Wong-Staal, Detection of lymphocytes expressing human T lymphotropic

virus type III in lymph nodes and peripheral blood from infected individuals by *in situ* hybridization. Proc. Natl. Acad. Sci. USA 83:772 (1986).

32. R.D. Schrier, J.A. McCutchan, J.C. Venable, J.A. Nelson, and C.A. Wiley, T-cell induced expression of HIV in macrophages. J. Virol. 64:3280 (1990).
33. H.E. Gendelman, O. Naragan, S. Kennedy-Stoskopf, P.G.E. Kennedy, Z. Ghotbi, J.E. Clements, J. Stanley, and G. Pezeshkpour, Tropism of sheep lentiviruses for monocytes: Susceptibility to infection and virus gene expression increase during maturation of monocytes to macrophages. J. Virol. 58:67 (1986).
34. F. Gray, R. Gherardi, M. Bandrimont, et al., Leukoencephalopathy with multinucleated giant cells containing human immune deficiency virus-like particles and multiple opportunistic infections in one patient with AIDS. Acta Neuropath. (Berlin) 73:99 (1987).
35. C.A. Wiley, M. Grafe, C. Kennedy, and J.A. Nelson, Human immunodeficiency virus (HIV) and JC virus in acquired immune deficiency syndrome (AIDS) patients with progressive multifocal leukoencephalopathy. Acta Neuropathol. (Berlin) 76:338 (1988).
36. C.A. Wiley, and J.A. Nelson, Role of human immunodeficiency virus and cytomegalovirus in AIDS encephalitis. Am. J. Pathol. 133:73 (1988).
37. J.A. Nelson, P. Ghazal, and C.A. Wiley, Editorial review: Role of opportunistic viral infections in AIDS. AIDS 4:1 (1990).
38. R. Amand, F. Siegal, C. Reed, S. Forlenza, and J. Moore, Noncytocidal natural variants of human immunodeficiency virus isolated from AIDS patients with neurological disorders. Lancet ii:234 (1987).
39. D. Serwadda, N.K. Sewankambo, and J.W.Carswell, Slim disease: A new disease in Uganda and its association with HTLV-III infection. Lancet ii:849 (1985).
40. T.C. Quinn, J.M. Mann, J.W. Curran, and D. Piot, AIDS in Africa: An epidemiologic paradigm. Science 234:955 (1986).
41. R. Ulrich, M. Zertz, W. Hiese, M. Lagi, G. Hoffken, and E.O. Reicken, Small intestinal structure and function in patients infected with human immunodeficiency virus (HIV):Evidence for HIV-induced enteropathy. Ann. Int. Med. 111:15 (1989).
42. P.A. Batman, A.R.O. Miller, S.M. Forster, J.R.W. Harris, A.J. Pinching, and G.E. Griffin, Jejunal enteropathy associated with human immunodeficiency virus infection: Quantitative histology. J. Clin. Pathol. 42:275 (1988).
43. D.P. Kotler, S.C. Weaver, and J.A. Terzakis, Ultrastructural features of epithelial cell degeneration in rectal crypts of patients with AIDS. Am. J. Surg. Pathol. 10:531 (1986).

44. J.A. Nelson, C.A. Wiley, C. Reynolds-Kohler, C.E. Reese, W. Margaretten, and J.A. Levy, Human immunodeficiency virus detected in bowel epithelium from patients with gastrointestinal symptoms. Lancet i:259 (1988).
45. J.A. Levy, W. Margaretten, and J.A. Nelson, Detection of HIV in enterochromaffin cells in the rectal mucosa of an AIDS patient. Amer. J. Gastro. 84:787 (1989).
46. J.M. Mathis, M. Hong, J. Grierson, D.E. Dwyer, C. Goldschmidt, D.A. Cooper, and A.L. Cunningham, HIV infection of rectal mucosa. Lancet i:1111 (1988).
47. W.D. Selby, G. Janossy, M. Bofill, and D.P. Jewell, Intestinal lymphocyte subpopulations in inflammatory bowel disease: an analysis by immunohistological and cell isolation techniques. Gut 25:32 (1984).
48. M. Zeitz, W.C. Greene, N.J. Peffer, and S.D. James, Lymphocytes isolated from the intestinal lamina propria of normal non-human primates have increased expression of genes associated with T-cell activation. Gastroenterology 94:647 (1988).
49. M. Budhraja, H. Levendorhn, F. Kocka, M. Mangkornkanok, and R. Sheret, Duodenal mucosal T cell subpopulation and bacterial cultures in acquired immune deficiency syndrome. Amer. J. Gastro. 82:427 (1987).
50. V.D. Rodgus, R. Fassett, and M.F. Kagnoff, Abnormalities in intestinal mucosal T cells in homosexual populations including those with the lymphadenopathy syndrome and acquired immunodeficiency syndrome. Gastroenterology 90:552 (1986).
51. S. Ellankany, T.L. Whiteside, R.R. Schade, and D.H. Van Thiel, Analysis of intestinal lymphocyte subpopulations in patients with acquired immunodeficiency syndrome (AIDS) and AIDS-related complex. Am. J. Clin. Path. 87:356 (1987).
52. D.P. Kotler, J.V. Scholes, and A.R. Tierney, Intestinal plasma cell alterations in the acquired immunodeficiency syndrome. Dig. Dis. Sci. 32:129 (1987).
53. AIDS Weekly Surveillance Report, United States AIDS Program. Atlanta Public Health Service Centers for Infectious Diseases, Centers for Disease Control. February 2. (1987).
54. S. Blanche, C. Rouzioux, G. Moscato, and M.L. Moscato, A prospective study of infants born to women seropositive for human immunodeficiency virus type 2. N. Engl. J. Med. 320: 1643 (1989).
55. Editorial, Vertical Transmission of HIV. Lancet ii:1057 (1988).
56. R.W. Marion, A.A. Wiznia, and R.G. Hutcheon, Fetal AIDS syndrome score. Correlation between severity of dysmorphism and age at diagnosis of immunodeficiency. AJDC 141:429 (1988).

57. M. Perro, R.W. Marion, and A.A. Wiznia, Dysmorphism leading to the diagnosis of acquired immunodeficiency syndrome. AJDC 141:474 (1987).

58. Q.H. Qazi, T.M. Sheikh, and S. Fikrig, Lack of evidence for craniofacial dysmorphism in perinatal human immunodeficiency virus infection. J. Ped. 112:7 (1988).

59. N. La Pointe, J. Michaud, D. Perovic, J.P. Chausseau, and J-M. Dupuy, Correspondence: Transplacental transmission of HTLV-III virus. N. Engl. J. Med. 312:1325 (1985).

60. S. Spreeher, G. Soumenkoff, F. Piussant, and M. Degueldre, Vertical transmission of HIV in 15 week fetus. Lancet ii: 288 (1986).

61. E. Jovaisas, M.A. Koch, A. Schaafer, M. Stauber, and D. Lowenthal, LAV/HTLV III in 20 week fetus. Lancet ii:1129 (1985).

62. W. Maury, B.J. Potts, and A.B. Robson, HIV-1 infection of first trimester and term human placental tissue: a possible mode of maternal-fetal transmission. J. Infect. Dis. 160: 583 (1989).

63. B.L. Drake, and J.C. Rodger, Phagocytic properties of culture trophoblast. Placenta 8:129 (1987).

64. J. Goldstein, M. Braverman, C. Salafia, and P. Buckley, The phenotype of human placental macrophages and its variation with gestational age. Am. J. Pathol. 244:648 (1988).

65. R. Demis, and T. Erbengi, Some new findings about Hofbauer cells in chorionic villi of the human placenta. Acta Anat. 119:18 (1984).

# VIRUS-HOST CELL INTERACTIONS IN HUMAN IMMUNODEFICIENCY VIRUS INFECTIONS

Suzanne Gartner, Kunihiro Ohashi and Mikulas Popovic

Primate Research Institute, New Mexico State University

P.O. Box 1027 Holloman AFB, New Mexico 88330

## INTRODUCTION

Considerable evidence indicates that the CD4 molecule, which is expressed primarily on helper T lymphocytes and cells of the mononuclear phagocyte lineage, serves as an essential component of the cell surface receptor for both the human and simian immunodeficiency viruses.[1,2,3] Fc receptor-mediated enhancement of HIV infection of monocytoid cells and infection of these cells via the complement receptor have also reported.[4,5] The expression of CD4 molecules on the surface of CD4-transfected human fibroblastoid and epitheloid cells confers susceptibility to infection with certain isolates of HIV-1.[6] Yet, the in vitro infection of CD4-negative fibroblasts appears to depend on perhaps somewhat artifactual conditions such as the use of very high multiplicities of infection and the subsequent requirement of cocultivation with CD4+ cells to demonstrate the presence of detectable HIV.[7] More recently, it has been shown that IgG Fc receptors can be induced on the surface of fibroblasts by cytomegalovirus and that these cells become susceptible to immune complex-mediated HIV-1 infection as a consequence.[8] However, the significance of the infection of fibroblasts and other CD4 negative cells to the pathogenesis of AIDS remains unclear.

Given the importance of the functional interactions between the helper T cell and the macrophage, particularly antigen presentation, we evaluated, in vitro, HIV-1 transmission from monocyte/macrophages to unstimulated autologous T cells.[9] The efficiency of this cell-to-cell-mediated transmission was approximately threefold greater than that observed in cell-free transmission to the unstimulated lymphocytes. Antibodies to CD4, antibodies to MHC class II antigens and soluble immunopurified gp120 all blocked the cell-mediated transmission. Moreover, HIV-1 production in the lymphocytes was greater in those populations additionally stimulated by exogenous antigen, either tetanus toxoid or streptokinase. The findings suggest that the virus can be readily transmitted from macrophage to lymphocyte, particularly within the context of antigen presentation, and that activation of the lymphocyte is important for this transmission. It it quite conceivable, then, as certain clinical findings have previously intimated[10], that the profound depletion of CD4+ helper T cells in AIDS patients is, at least, partially a consequence of the elimination of antigen-specific T cell clones.

To further elucidate interactions between HIV-1, mononuclear phagocytes and T lymphocytes with potential relevance to disease development, we experimentally addressed the following issues: (1) the relationship between infection and the affinity of virus binding to the CD4 receptor, (2) the role of the CD4 receptor in HIV infection of mononuclear phagocytes, and (3) differences in host cell susceptibility among various HIV-1 isolates.

## MATERIALS AND METHODS

Binding studies. Optimal concentrations of the OKT4, OKT4A, OKT4D and OKT4F antibodies (Ortho) were determined by titration. High concentrations ($5x10^{10}$ virus particles per $10^6$ cells) of sucrose gradient-purified viruses were used. H9 cells were exposed to the virus preparations for 10, 30, 60 or 120 minutes, then washed with phosphate buffered saline without calcium and magnesium (PBS), stained with the monoclonal antibodies using standard procedures[11] and analyzed using flow cytometry.

Blocking with monoclonal antibodies. Optimal concentrations of the OKT4, OKT4A, OKT4D and OKT4F antibodies were determined by titration. OKT3 was used as a negative control antibody. $4x10^6$ adherence-purified monocytes were pre-treated with the monoclonal antibody for 30 minutes at 37° C. The cell suspension was then diluted with PBS, centrifuged and the cell pellet resuspended in 1ml of the virus inoculum containing 50,000 cpm of reverse transcriptase (RT) activity, and incubated at 37° C. for 1 hour. The cells were then washed with 10ml of complete medium (RPMI 1640 supplemented with 20% fetal bovine serum, 2mM L-glutamine and 100 units/ml of penicillin and 100 mcg/ml streptomycin), and seeded in T25 flasks in complete medium. Virus expression was monitored by RT assay.

Blocking with gp120 and gp160. Fifty microliters of freshly-isolated monocyte/macrophage cell suspension ($2x10^7$ cells/ml) was mixed with 50 microliters of varying concentrations of HIV-1 (HTLV-$III_{451}$)[12] gp120 or gp160, and incubated for 45 minutes at 37° C. The cells were then washed and incubated in 0.1ml of the HIV-1 (HTLV-$III_{Ba-L}$) viral inoculum containing 5000 cpm/ml RT activity for 45 minutes at 37° C. Following infection, the cells were washed with complete medium, resuspended in 0.6ml of complete medium additionally supplemented with 10% human serum, plated into 6 microtiter wells, 0.1ml per well, and incubated at 37° C in 5% $CO_2$ in air. Virus expression was monitored by p24 antigen capture assay (Dupont).

Immunoprecipitation of complexes. Purified HTLV-$III_{451}$ gp120 was labeled with $^{125}I$ in the presence of Iodogen to a specific activity of 2000 cpm/ng. T75 flasks of nearly confluent monolayers of human monocyte/macrophages were incubated at room temperature for two hours with 200,000 cpm of the $^{125}I$-labeled gp120 in 5ml of RPMI 1640 medium supplemented with 20mM Hepes buffer pH 7.2 and 2mg/ml bovine serum albumin. The medium was then aspirated off, the cells were washed three times with PBS and solubilized in 3 ml of PBS containing 1% triton X-100, 0.5% deoxycholate and 0.1mM phenylmethylsulfonyl fluoride. The extract was centrifuged at 30,000rpm for 30 minutes and 1ml of the sample was immunoprecipitated with either OKT4, OKT4A or goat anti-gp120 antiserum as previously described.[12] The immunoprecipitates were run in 7.5% polyacrylamide gels and visualized by autoradiography. Alternatively, metabolic labeling of cells was carried out as previously described.[12] Cultures of monocyte/macrophages were incubated for 15 hours in the presence of $^{35}S$-labeled methionine.

Cell culture and virus infections. Peripheral blood-derived monocyte/macrophages were cultured as previously described.[13] Human and chimpanzee T lymphocyte cultures were established using standard methods.[14] Peripheral blood mononuclear cells were stimulated for three days with PHA-P and then maintained in the presence of Interleukin 2. For the comparative infections of human and chimpanzee T cells, cell-free viral inocula containing between 500,000 and 700,000 cpm of reverse transcriptase (RT) activity were used. Infections were carried out as previously described.[13] Both cell-free virus infections and cocultivations were performed in the infectivity studies of normal human and transformed cells. Serial tenfold dilutions of the cell-free viral inocula were used, the highest dilutions containing between 500,000 and 1,000,000 cpm RT activity/ml. Cocultivations were performed with either irradiated (7000 rads) neoplastic T cell producers of HTLV-$III_B$ or normal infected monocyte/macrophages. The normal mesothelial cells were derived from pleural effusions. The T lymphoblast cell lines H9,

Molt-3 and CEM, the B lymphoblast cell lines P3HR and LB-Brown, and the astrocytoma cell line HG-66-85[15] were used. The infections were carried out as previously described.[13] The osteogenic sarcoma cell line, HOS, was transfected with a molecular construct containing the gene encoding the CD4 molecule, and shown to express high cell surface levels of the protein.

## RESULTS

Table 1. Binding of HTLV-$III_B$ and HTLV-$III_{RF}$ to CD4 Epitopes.

| Isolate/Moabs | MFU of CD4 epitopes after exposure to the virus ( time in minutes) | | | | |
|---|---|---|---|---|---|
| | 0 | 10 | 30 | 60 | 120 |
| HTLV-$III_B$ | | | | | |
| OKT4A | 430 | 190 | 45 | 10 | 0 |
| OKT4D | 424 | 191 | 65 | 11 | 0 |
| OKT4F | 400 | 40 | 10 | 5 | 0 |
| OKT4 | 651 | 659 | 532 | 497 | 413 |
| HTLV-$III_{RF}$ | | | | | |
| OKT4A | 890 | 660 | 320 | 240 | 200 |
| OKT4D | 700 | 450 | 300 | 200 | 150 |
| OKT4F | 330 | 250 | 200 | 150 | 140 |
| OKT4 | 880 | 890 | 890 | 720 | 680 |

Moabs = monoclonal antibodies directed against different epitopes of the CD4 molecule.
MFU = mean fluorescence units determined by flow cytometry.

Affinity of virus binding. Experiments were performed using H9 cells to compare the binding affinity of HTLV-$III_B$ and HTLV-$III_{RF}$ for epitopes of the CD4 molecule. Blocking of the binding of specific monoclonal antibodies to the CD4 epitopes was taken as an indirect measurement of HIV-1 binding. The HTLV-$III_{RF}$ isolate was chosen for comparison because it was adapted to continuous production in the H9 cell line at about the same time as the HTLV-$III_B$ isolate and it can readily be produced for concentration in quantities comparable to HTLV-$III_B$. Unlike HTLV-$III_B$, however, HTLV-$III_{RF}$ can more efficiently infect peripheral blood-derived monocyte/macrophages and it exhibits a more pronounced cytopathic effect in normal T cells and monocyte/macrophages (S. Gartner, M. Popovic and E. Read-Connole, unpublished data). As shown in Table

1, the blocking of the CD4 A, D and F epitopes by HTLV-III$_B$ was significantly greater than that observed with HTLV-III$_{RF}$. For example, after 30 minutes of exposure to the virus, 90% of the CD4A sites were occupied in the case of HTLV-III$_B$ compared to only 64% in the case of HTLV-III$_{RF}$. Similarly, essentially 100% of the OKT4F epitopes were blocked by a 10 minute exposure to HTLV-III$_B$ compared to 25% for HTLV-III$_{RF}$. Also, the kinetics of virus binding as determined by time-dependent blocking of the epitopes suggests that HTLV-III$_B$ has a greater affinity for the CD4 receptor. In addition, recent preliminary results suggest that the HTLV-III$_{Ba-L}$ isolate used in the studies described here has a lower affinity for the CD4 receptor than HTLV-III$_B$ (M. Rosenberg, personal communication).

Table 2. Evidence for CD4-Mediated HIV-1 Infection of Monocyte/macrophages in vitro.

| Blocking of monocyte/macrophage infection with HTLV-III$_{Ba-L}$ procedure | results |
|---|---|
| exposure of cells to moabs (OKT4A, OKT4D, OKT4F) directed against CD4 prior to virus infection | 4 to 10-fold decrease in viral yield |
| exposure of cells to HIV-1 gp120 prior to virus infection | 12.5ug/ml of gp120 reduced virus infection by 50% |
| **Detection of CD4--HIV-1 env complexes** | |
| precipitation of $^{35}$S-methionine-labeled gp120/160 from cells with | detected complexes |
| a) goat anti-gp120 antiserum | gp120/160--CD4 |
| b) OKT4 | gp120/160--CD4 |
| c) OKT4A | none |
| precipitation of $^{125}$I-labeled gp120 from cells with | |
| a) goat anti-gp120 antiserum | gp120--CD4 |
| b) OKT4 | gp120--CD4 |
| c) OKT4A | none |

CD4 and the infection of monocyte/macrophages. While it seems likely that the human immunodeficiency viruses utilize the CD4 molecule as a means of entry into mononuclear phagocytes, more definitive proof is needed. To this end, we carried out experiments to determine if the infection of these cells could be blocked by pretreatment with anti-CD4 monoclonal antibodies and if complexes between the viral envelope glycoprotein and the CD4 molecule could be precipitated from the virus-exposed cell populations. A summary of these results are shown in Table 2. Monoclonal antibodies directed against the A, D and F epitopes of the CD4 molecule each blocked HIV-1 infection of cultured, peripheral blood-derived, monocyte/macrophages (M/M). Moreover, pretreatment of M/M with HIV-1 gp120 prior to virus exposure resulted in a 50% reduction in virus infection. As Table 2 also notes, precipitable complexes between CD4 and HIV-1 gp120 and gp120/160 and were detected using and $^{125}I$-labeled gp120 and $^{35}S$ metabolic labeling, respectively.

Table 3. Replication of Various HIV-1 Isolates in Human Cells of Different Lineages.

| | virus isolate | | | |
|---|---|---|---|---|
| cell type | $III_B$ | Ba-L | RC-br | WR-sk |
| Normal | | | | |
| helper T lymphocyte | + | + | + | + |
| monocyte/macrophage | + | + | + | + |
| skin fibroblast | - | - | - | - |
| mesothelial cell | - | - | - | ND |
| Transformed | | | | |
| CD4+ T lymphoblast | + | + | - | - |
| CD4+ monocytoid cell | + | + | - | - |
| EBV* B lymphoblast | + | - | - | ND |
| astrocytoma cell | + | - | - | - |
| CD4+ sarcoma cell** | + | - | - | - |

Virus replication was determined by reverse transcriptase assay.
*EBV - Epstein-Barr virus-immortalized B lymphocyte.
**This cell line was developed by transfection with a molecular construct containing the CD4 receptor gene.

Host cell susceptibility. The genomic heterogeneity of HIV-1 has been well-documented.[16] Biological variability has also been observed, most notably differences in replicative potential, transmissibility and host cell tropism.[13,17] Controversy exists, however, regarding the host cell range of the virus. We have observed that the HTLV-$III_B$ isolate manifests properties different from those common to fresh, field isolates of the virus. To evaluate further the susceptibility of different kinds of cells to the virus, we performed a series of in vitro infections using both normal and transformed cells. The isolates HTLV-$III_{Ba-L}$, HTLV-$III_{RC-br}$, and HTLV-$III_{WR-sk}$, isolated from lung,[13] brain,[18] and skin[19] tissue, respectively, were propagated only in normal T cells and monocyte/macrophages prior to use. As shown in Table 3, all isolates were able to productively infect normal peripheral blood-derived T helper cells and monocyte/macrophages. In contrast, only HTLV-$III_B$ and HTLV-$III_{Ba-L}$ were able to infect the neoplastic T lymphoid (H9 and Molt-3) and monocytoid (U-937) cell lines, and only HTLV-$III_B$ was also able to infect EBV-transformed B cells as well as neoplastic glial cells[15] and osteogenic sarcoma cells bearing CD4 receptors. High multiplicities of infection were required (~1,000,000 cpm RT activity) for HTLV-$III_{Ba-L}$ to infect the neoplastic cells. Moreover, cocultivation was required for HTLV-$III_B$ to infect the astrocytoma cells.

The unique biological behavior of HTLV-$III_B$ was also revealed in experiments comparing the susceptibility of chimpanzee and human cells to different virus isolates. Persistent in vivo infection of chimpanzees with the HTLV-$III_B$ isolate has previously been reported.[20] As shown in Table 4, cultures of mitogen-stimulated T lymphocytes from all five chimpanzees were susceptible to infection with the HTLV-$III_B$ isolate, but not with the HTLV-$III_{Ba-L}$ and HTLV-$III_{WR-sk}$ isolates. In addition, T lymphocytes from the two animals evaluated were also susceptible to infection with the HIV-$1_{LW-PBM}$ isolate which was recovered from a laboratory worker shown to be infected with the HTLV-$III_B$ isolate.[21] In contrast, all cultures of human T lymphocytes were susceptible to all four of the HIV-1 isolates tested.

Table 4. Susceptibility of Chimpanzee and Human T Lymphocytes to HIV-1 Isolates as Determined by RT assay.

| | | virus isolate | | | |
|---|---|---|---|---|---|
| species | specimen | $III_B$ | LW-PBM | Ba-L | WR-sk |
| human | buffy coat #630 | + | + | + | + |
| | buffy coat #7-10 | + | + | + | + |
| | buffy coat #6615 | + | + | + | + |
| chimpanzee | #258 | + | + | - | - |
| | #938 | + | + | - | - |
| | #184 | + | ND | - | - |
| | #1315 | + | ND | - | - |
| | #192 | + | ND | - | - |

## DISCUSSION

The evidence presented indicates that HIV-1 utilizes the CD4 molecule present on the surface of monocyte/macrophages, as well as T helper lymphocytes, to enter the target cell. While other routes of entry are obvious possibilities, especially considering certain properties of macrophages (i.e., Fc receptors and phagocytic activity), the degree of inhibition of infection observed in these studies suggest that specific receptor mediated-entry is probably the most efficient and, therefore, most common mode.

The H9/HTLV-$III_B$ virus-cell system is used by large numbers of investigators to study many different aspects of HIV-1. This system offers certain advantages, particularly in terms of virus propagation and large-scale purification of viral components, and studies of cytopathicity. It also readily lends itself to molecular analyses and structure-function studies. We previously reported that HTLV-$III_B$ differs from field isolates in having lost tropism for cells of the mononuclear phagocyte lineage.[13] We report here that HTLV-$III_B$ also differs from field isolates in other ways. HTLV-$III_B$ was the only isolate evaluated that was able to productively infect Epstein-Barr virus-driven B lymphoblasts, an astrocytoma cell line and an osteogenic sarcoma cell line bearing CD4 receptor molecules. The factors responsible for these unique behaviors of the HTLV-$III_B$ isolate are not known. It is possible that the adaptation of the virus to continuous production in the neoplastic cell played a role, especially since it is the general experience among those performing HIV-1 isolation that only a few isolates of the virus readily infect neoplastic T cells. Data presented here demonstrate that HTLV-$III_B$ has a significantly higher affinity for the CD4 receptor molecule than does HTLV-$III_{RF}$, another isolate adapted to continuous production in the neoplastic cell. Possibly, this greater binding affinity of HTLV-$III_B$ can, at least, partially account for its extended host cell range with respect to cells expressing low levels of the CD4 molecule and cells genetically engineered to express CD4. Similarly, the ability of HTLV-$III_B$, but not the other isolates, to infect the chimpanzee T cells may relate to binding affinity. The observed differences in binding affinity may reflect subtle differences in the interactions between the virus envelope protein and the binding epitopes of the CD4 molecule. The explanations for differences between HTLV-$III_B$ and other HIV-1 isolates will hopefully be forthcoming. The important point to be made here is that generalizations regarding the behavior of HIV-1 should not be made solely on the basis of studies performed using HTLV-$III_B$ or the H9/HTLV-$III_B$ system or, most likely, on the basis of any similar virus-cell system. Also, recognizing that only certain HIV-1 isolates can infect chimpanzee cells and determining the reasons why, is of critical importance, since endstage vaccine testing will presumably be carried out in these animals, and consequently, it will be necessary to have in hand, the most appropriate challenge viruses.

SUMMARY

The isolate HTLV-$III_B$ is used by a large number of investigators for a variety of studies. The data shown here emphasize the uniqueness of this isolate with respect to certain biological behaviors, in particular, the ability to infect cells other than T lymphocytes and mononuclear phagocytes, and the ability to infect chimpanzee T cells. Data also presented indicates that HTLV-$III_B$ has a greater binding affinity for the CD4 receptor molecule, which may help to explain its observed biological uniqueness. Caution in generalizing from findings based on studies with HTLV-$III_B$ alone is recommended. In addition, evidence demonstrating CD4-mediated entry of HIV-1 into normal human monocyte/macrophages is included.

ACKNOWLEDGMENTS

We wish to thank Dr. Dean L. Mann for his participation in the CD4 epitope binding studies and Dr. Anna Aldovini for the CD4-transfected human osteogenic sarcoma cell line. Some of these studies were conducted in the Laboratory of Tumor Cell Biology, National Cancer Institute, National Institutes of Health.

REFERENCES

1. Popovic, M., Read-Connole, E. and Gallo, R.C., T4 positive human neoplastic cell lines susceptible to and permissive for HTLV-III. Lancet ii:1472 (1984).
2. Klatzmann, D., Champagne, E., Chamaret, S., Gruest, J., Guetard, D., Hercend, T., Gluckman, J.-C. and Montagnier, L., T-lymphocyte T4 molecule behaves as the receptor for human retrovirus LAV. Nature 312:767 (1984).
3. Dalgleish, A.G., Beverley, P.C.L., Clapham, P.R., Crawford, D.H., Greaves, M.F. and Weiss, R.A., The CD4 (T4) antigen is an essential component of the receptor for the AIDS retrovirus. Nature 312:763 (1984).
4. Takeda, A., Tuazon, C.U. and Ennis, F.A., Antibody-enhanced infection by HIV-1 via Fc receptor-mediated entry. Science 242:580 (1988).
5. Robinson Jr., W.E., Montefiori, D.C. and Mitchell W.M., Antibody-dependent enhancement of human immunodeficiency virus type 1 infection. Lancet i:790 (1988).

6. Chesebro, B.and Wehrly, K., Development of a Sensitive Quantitative Focal Assay for Human Immunodeficiency Virus Infectivity. J. Virol. 62(10):3779 (1988).
7. Tateno, M., Gonzalez-Scarano, F. and Levy J.A., Human immunodeficiency virus can infect CD4-negative human fibroblastoid cells. Proc. Natl. Acad. Sci. USA 86:4287 (1989).
8. McKeating, J.A., Griffiths, P.D. and Weiss, R.A., HIV susceptibility conferred to human fibroblasts by cytomegalovirus-induced Fc receptor. Nature 343:659 (1990).
9. Mann, D.L., Gartner, S., LeSane, F., Buchow, H. and Popovic, M. HIV-1 transmission and function of virus-infected monocytes/macrophages. J. Immunol. 144:2152 (1990).
10. Lane, H.C. and Fauci, A.S., Immunologic abnormalities in the acquired immunodeficiency syndrome. Ann. Rev. Immunol. 3:477 (1985).
11. Mann, D.L., Popovic, M., Murray, C.H., Neuland, C., Strong, D.M., Sarin, P.S., Haynes, F.B., Gallo, R.C., and Blattner, W.A., Cell surface antigen expression in newborn cord blood lymphocytes infected with HTLV. J. Immunol. 131:2021 (1983).
12. Kalyanaraman, V.S., Pal, R., Gallo, R.C. and Sarngadharan, M.G. A unique human immunodeficiency virus culture secreting soluble gp160. AIDS Res. and Human Retro. 4(5):319 (1988).
13. Gartner, S., Markovits, P., Markovitz, D.M., Kaplan, M.H., Gallo, R.C. and Popovic, M., The role of mononuclear phagocytes in HTLV-III/LAV infection. Science 233:215 (1986).
14. Morgan, D.A., Ruscetti, F.W. and Gallo, R.C., Selective in vitro growth of T-lymphocytes from normal human bone marrow. Science 193:1007 (1976).
15. Popovic, M., Mellert, W. and Gartner, S., Role of mononuclear phagocytes and accessory cells in HIV-1 infection of the brain. Ann. Neurol. (Suppl) 23:574 (1988).
16. Wong-Staal, F., Shaw, G.M., Hahn, B.H., Salahuddin, S.Z., Popovic, M., Markham, P., Redfield, R. and Gallo, R.C. Genomic diversity of human T-lymphotropic virus type III (HTLV-III). Science 229:759 (1985).
17. Asjo, B., Albert, J., Karlsson, A., Morfeldt-Manson, L., Biberfeld, G., Lidman, K. and Fenyo, E.-M. Replicative capacity of human immunodeficiency virus from patients with varying severity of HIV infection. Lancet ii:660 (1986).
18. Gartner, S., Markovits, P., Markovitz, D.M., Betts, R.F. and Popovic, M., Virus isolation from and identification of HTLV-III/LAV-producing cells in brain tissue from an AIDS patient. JAMA 256:2365 (1986).
19. Rappersberger, K., Gartner, S., Schenk, P., Stingl, G., Groh, V., Tschachler, E., Mann, D.L., Wolff, K., Konrad, K. and Popovic, M. Langerhans cells are an actual site of HIV-1 replication. Intervirol. 29:185 (1988).

20. Nara, P.L., Robey, W.G., Arthur, L.O., Asher, D.M., Wolff, A.V., Gibbs Jr., C.J., Gajdusek, D.C. and Fischinger, P.J. Persistent infection of chimpanzees with human immunodeficiency virus: serological responses and properties of reisolated viruses. J. Virol. 61(10):3173 (1987).
21. Weiss, H.S., Goedert, J., Gartner, S., Popovic, M., Waters. D., Markham, P.D., Veronese, F., Gail, M., Barkley, W.E., Gibbons, J., Gill, F.A., Leuther, M. Shaw, G.M., Gallo, R.C. and Blattner, W.A. Risk of human immunodeficiency virus (HIV-1) infection among laboratory workers. Science 239:68 (1988).

# MEDIATORS OF FUSION BETWEEN HIV-INFECTED MACROPHAGES AND LYMPHOID CELLS

Suzanne Crowe[1], John Mills[2,3] and
Michael S. McGrath[2,4]

[1]MacFarlane Burnet Centre for Medical Research, Fairfield, Victoria, Australia; [2]Medical Service, San Francisco General Hospital; [3]Departments of Medicine, Microbiology and Laboratory Medicine, University of California, San Francisco. Corresponding author; [4]Department of Medicine and Laboratory Medicine, University of California, San Francisco

Shortly after the discovery of the human immunodeficiency virus (HIV) and recognition that it was the causative agent of the acquired immunodeficiency syndrome (AIDS), the CD4 molecule was identified as the principal receptor for the virus by Dalgleish and Klatzmann and their collaborators (1,2). The CD4 molecule is a member of the immunoglobulin superfamily, and was originally recognized as a surface marker for the helper-inducer T lymphocyte subset (3). However, other cells are known to express CD4, including follicular dendritic cells and cells of the monocyte-macrophage series (4,5,6). It was thus not surprising that by 1987 several groups had shown that monocyte-macrophages were susceptible to HIV infection (6-8). Infection of normal monocyte-macrophages cultured <u>in vitro</u> with laboratory strains of HIV resulted either in no infection whatsoever (as is characteristic of the HTLV-IIIB strain of HIV (6,8)), or in infection of up to 70% of the cultured cells (6-8). HIV infection of monocyte macrophages has very

*Mechanisms and Specificity of HIV Entry into Host Cells*
Edited by N. Düzgüneş, Plenum Press, New York, 1991

different characteristics than infection of peripheral blood lymphocytes or lymphoid cells. Replication of the virus was comparatively delayed, with peak titers of virus and viral products being reached many days or weeks after infection (6). Although some viral cytopathology was induced, primarily multinucleate giant cell formation, the infected cells survived in culture and continued to produce virus for periods of up to several months (6-8,9). These properties of the in vitro infection have led several investigators to hypothesize that long-lived tissue macrophages might serve as a reservoir for HIV infection in vivo (6,9). Certainly a reservoir of virus replication other than CD4 lymphocytes must exist, as recent studies have shown that patients with HIV infection develop progressively increasing titers of virus in plasma with time, as the CD4 lymphocyte count falls inexorably toward virtually undetectable levels (10,11). Tissue monocyte-macrophages are certainly a possible source for this virus, as advanced HIV infection is not associated with marked reduction in circulating or tissue monocyte-macrophage numbers (12 and Crowe SM, unpublished data).

Lifson and collaborators noted fusion between HIV-infected lymphocytes and VB cells, a lymphoblastoid cell line expressing large amounts of surface CD4 (13). If fusion were to occur between HIV infected monocyte-macrophages and lymphocytes, it might provide for a mechanism by which a tissue reservoir of chronically-infected monocyte-macrophages might remove CD4-bearing lymphocytes; it might also offer an alternative route of infection for those cells. We therefore looked for fusion between cultured monocyte-macrophages, infected with HIV, and uninfected CD4-bearing lymphoid cells. As shown in Figure 1, addition of cells with surface CD4 to infected monocyte-macrophages resulted in the formation of multinucleate giant cells within a day or two; giant cell formation did not occur if the added cells were isogeneic but CD4 negative. The time course of giant cell formation -- within 1-2 days -- was inconsistent with the effect being due to cell-free virus, as addition of cell-free virus to VB cells did not result in fusion until after 2 days of incubation. However, to control for this possibility, we used a technique similar to that described by Lifson and his collaborators (13). Cultured macrophages that had been previously infected with HIV were labeled with fluorescein, while uninfected VB cells were labeled with rhodamine (13). The labelled cells were mixed, and all giant cells were examined under a fluorescent

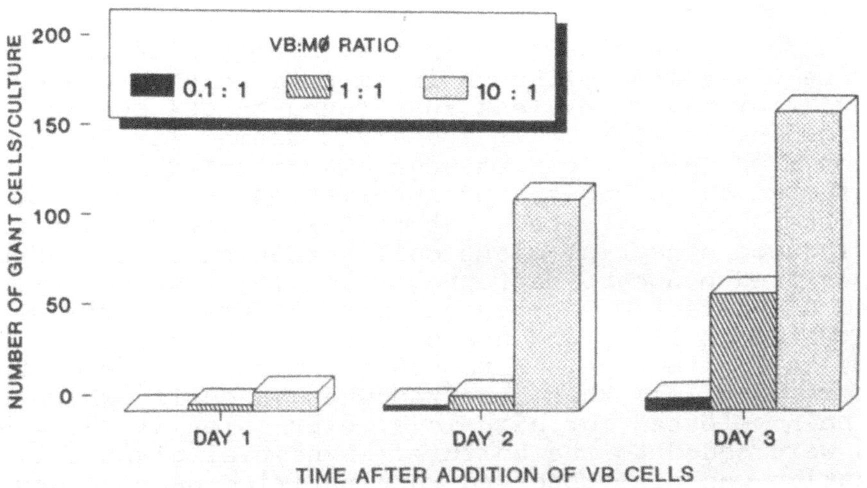


Figure 1. Time course of fusion between HIV-infected, cultured monocyte macrophages and CD4-bearing lymphoid cells (VB cells). Peripheral blood monocytes were separated from buffy coats of normal, HIV-seronegative blood donors by Ficoll-Hypaque density centrifugation followed by a single cycle of adherence purification. The resulting cells, which were 85-95% monocytes, were cultured in suspension in perfluoralkoxy (equivalent to Teflon) jars. The cultured macrophages were infected with the DV strain of HIV, and synthesis of cell-associated HIV p24 antigen monitored cytofluorographically (6). After 10-14 days, cultures in which >25% of the macrophages were expressing p24 antigen were transferred to multichamber microscope culture-slides (Labtek), $10^5$ macrophages/chamber, and VB cells added at a 10:1 ratio (VB:macrophage). The presence of multinucleated giant cells was scored by daily observation.

microscope equipped with filters selective for both rhodamine and fluorescein. All syncytial giant cells examined contained both rhodamine and fluorescein, while individual VB cells and macrophages contained only the original dye (13). The bifluorescent cells were not the result of phagocytosis, as the time course for fusion (1-2 days; Figure 1) was far longer than required for phagocytosis, and bifluorescent cells were not observed when **un**infected macrophages were mixed with VB cells. Fusion was also confirmed by electron microscopy (Fig. 2).

We and others have shown that monocyte-macrophages express CD4 (5,6), although in amounts which are about an order of magnitude less than found on CD4-bearing lymphoid cells (5,6). It therefore seemed likely that fusion might also occur between HIV-infected and uninfected macrophages, and observations by ourselves and others had suggested that at least some strains of HIV induced syncytial giant cell formation in cultures of purified monocyte-macrophages (6,8,9). We therefore mixed HIV-infected macrophages with uninfected cells, in a ratio of 1:10, and again looked for syncytial giant cell formation. Although giant cells formed, the numbers were less than observed when VB cells, which had been selected for high-level expression of surface CD4, were added to the culture. Syncytial giant cell formation was also observed when purified macrophages were added to HXB cells, a cell line chronically infected with the $HTLVIII_B$ strain of HIV (Crowe SM et al, unpublished data).

If fusion between chronically-infected monocyte macrophages and CD4-bearing lymphocytes is an important mechanism for depletion of CD4 lymphocytes or transmission of virus, it is of great importance to understand the surface molecules involved in the fusion mechanism. Based on the previous data of Lifson et al (13), and the fact that fusion was not observed between VB cells selected for absence of CD4, we hypothesized that fusion required an interaction between surface gp120 on the infected macrophages, and CD4 on the lymphoid cells. Such a mechanism conflicts with some previous reports of HIV infection of macrophages, which suggest that virus particles bud internally without resulting in surface expression of viral antigens (9). To address this question we examined the effect of a

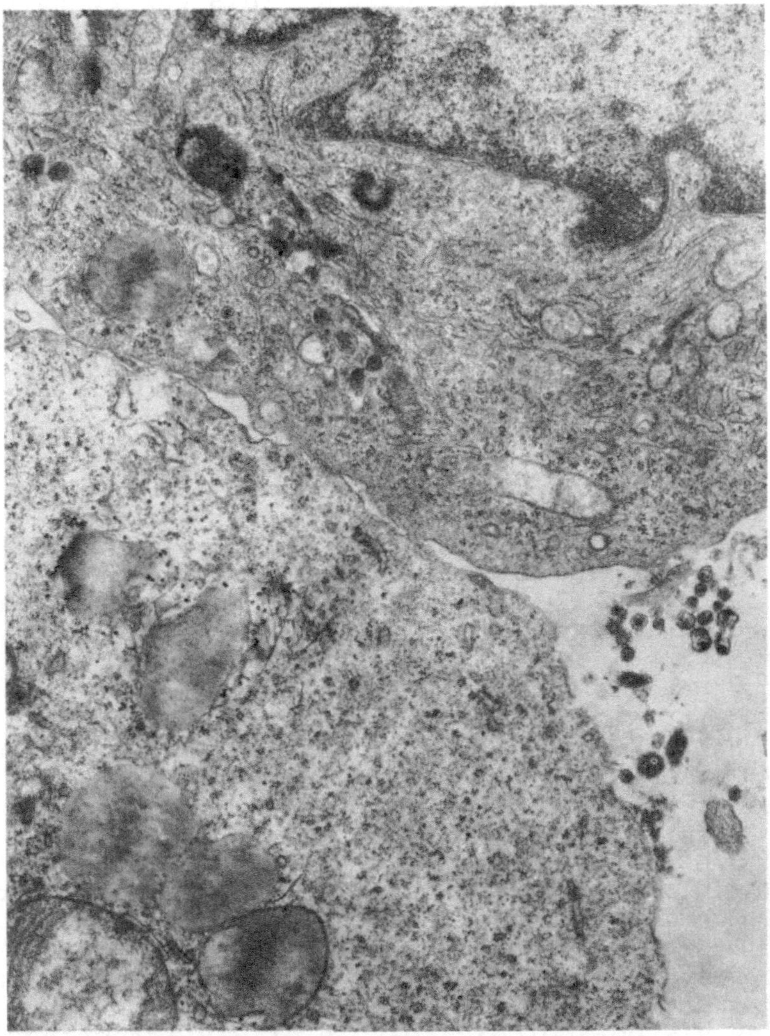

Figure 2. Fusion between HIV-infected macrophages and CD4-positive VB cells. A typical section showing fusion between macrophages (which also show budding virus particles) and VB cells is shown (original magnification 11,000x).

number of specific inhibitors on fusion between HIV-infected macrophages and VB cells (Figure 3). Fusion was inhibited in a dose-dependent fashion by recombinant, soluble CD4, confirming our previous data (using CD4-negative VB cells) that surface CD4 was required for the fusion reaction. This was further confirmed by studies with a monoclonal antibody to CD4, similar to Leu3a, that also inhibited the fusion reaction by >95% at $\geq$ 10 ug/ml (S3.5, a generous gift of Edward Engleman, Stanford University School of Medicine). Fusion was not inhibited by equal concentrations of an irrelevant monoclonal antibody of the same class and isotype as S3.5, MOPC-21.

The most likely ligand for CD4 is of course HIV gp120 (14), and the importance of gp120 in the fusion reaction was confirmed by showing that the fusion reaction was inhibited in a dose-dependent fashion by recombinant, fully-glycosylated gp120 (Figure 3). Fusion was not inhibited by a non-glycosylated N-terminal gp120 peptide, even though this material reacted with sera from patients with HIV infection (15).

Although both gp120 and CD4 are essential for fusion to occur, it is possible that other cellular adhesion molecules are also required for the reaction to occur. This has been shown for fusion of HIV-infected lymphoid cells, where antibody to LFA-1 inhibits the reaction (16). We therefore tested monoclonal antibodies to the integrins LFA-1, LFA-3 and ICAM-1 in the fusion reaction. Fusion was inhibited >95% by anti-LFA-1 at 20 ug/ml, but anti-LFA-3 and anti-ICAM-1 had no effect.

In conclusion, these experiments have documented that fusion can occur between HIV-infected macrophages and CD4-bearing lymphoid cells. These data support the role of macrophages as a potential scavenger of CD4 lymphocytes, through formation of multinucleate giant cells. Although syncytial giant cells are not a prominent histopathologic feature of HIV infection, they are observed regularly in the brain (17), and there are numerous reasons why they might not be observed in other tissues. These data also provide

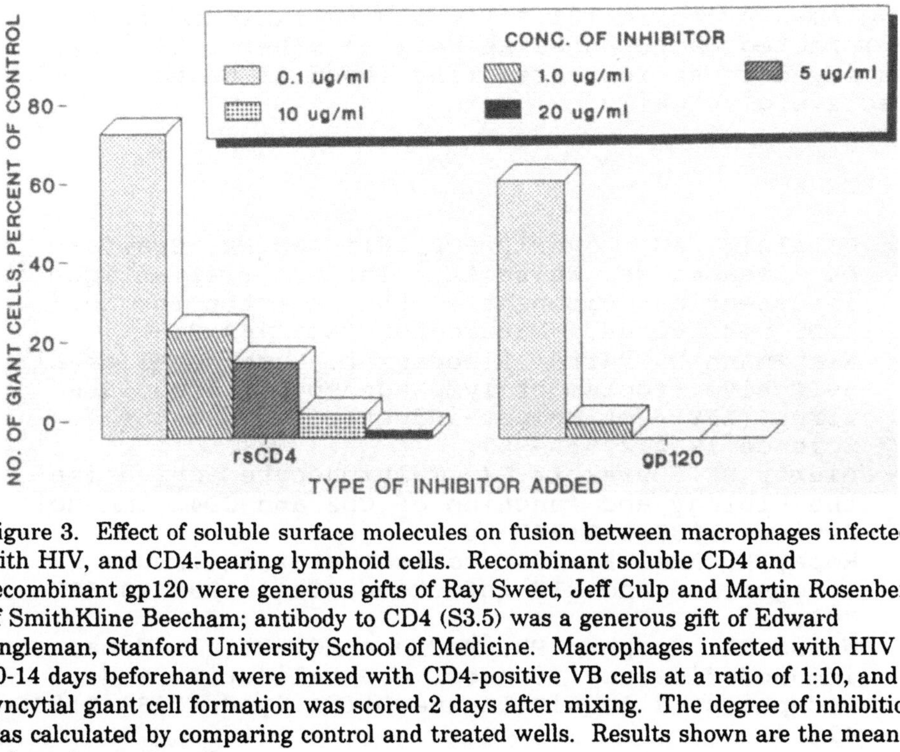


Figure 3. Effect of soluble surface molecules on fusion between macrophages infected with HIV, and CD4-bearing lymphoid cells. Recombinant soluble CD4 and recombinant gp120 were generous gifts of Ray Sweet, Jeff Culp and Martin Rosenberg of SmithKline Beecham; antibody to CD4 (S3.5) was a generous gift of Edward Engleman, Stanford University School of Medicine. Macrophages infected with HIV 10-14 days beforehand were mixed with CD4-positive VB cells at a ratio of 1:10, and syncytial giant cell formation was scored 2 days after mixing. The degree of inhibition was calculated by comparing control and treated wells. Results shown are the mean of at least 3 experiments.

strong evidence for the presence of surface gp120 on HIV infected macrophages. Our findings have considerable relevance to the development of chemotherapeutic regimens for HIV infection. Several agents are in clinical trials which depend on recombinant, soluble CD4 (18-20). Administration of such drugs would be expected to reduce fusion between macrophages (and other HIV-infected cells) and uninfected, CD4-bearing cells. This should reduce cell-to-cell transmission of virus, retarding the progression of the infection. Additionally, CD4 conjugated to toxins such as ricin or Pseudomonas aeruginosa exotoxin (21,22) would be expected to target HIV infected macrophages as well as other cells, and thus may have an increased likelihood of being effective clinically.

## REFERENCES


1. Dalgleish AG, Beverly PCL, Clapham PR, Crawford DH, Greaves MF, Weiss RA. The CD4 (T4) antigen is an essential component of the receptor for the AIDS retrovirus. Nature 1984;**312**:763-7.
2. Klatzmann D, Barre'-Sinoussi F, Nugeyre MT et al. Selective tropism of lymphadenopathy associated virus (LAV) for helper-inducter lymphocytes. Science 1984;**225**:59-63.
3. Bierer BE, Burakoff SJ. T-lymphocyte activation: the biology and function of CD2 and CD4. Immunol Reviews 1989;**111**:267-94.
4. Robey E, Axel R. CD4: collaborator in immune recognition and HIV infection. Cell 1990;**60**:697-700.
5. Weiss RA, Clapham PR, McClure MO, et al. Human immunodeficiency viruses: neutralization and receptors. J Acquir Immune Defic Syndr 1988;**1**:536-41.
6. Crowe SM, Mills J, McGrath MS. Quantitative immunocytofluorographic analysis of CD4 surface antigen expression and HIV infection of human peripheral blood monocyte-macrophages. AIDS Res Human Retroviruses 1987;**3**:135-45.
7. Gartner S, Markovits P, Markovitz DM, Kaplan MH, Gallo RC, Popovic M. The role of mononuclear phagocytes in HTLVIII/LAV infection. Science 1986;**233**:215-219.

8. Ho DD, Rota TR, Hirsch MS. Infection of monocyte-macrophages by human T lymphotropic virus type III. J Clin Invest 1986;**77**:1712-1715.
9. Gendelman HE, Orenstein JM, Martin MA, Ferrua C, Mitra R, Phipps T, Wahl LA, Lane HC, Fauci AS, Burke DS, Skillman J, Meltzer MS. Efficient propagation of HIV on recombinant colony-stimulating factor treated monocytes. J Exp Med 1988;**167**:1428-1441.
10. Ho DD, Moudgil T, Alam M. Quantitation of human immunodeficiency virus type 1 in the blood of infected persons. NEJM 1989;**321**:1621-5.
11. Coombs RW, Collier AC, Allain JP, et al: Plasma viremia in human immunodeficiency virus infection. NEJM 1989;**321**:1626-31.
12. Smith PD, Ohura K, Masur H, et al. Monocyte function in the acquired immune deficiency syndrome: defective chemotaxis. J Clin Invest 1984:**74**;2121-2128.
13. Lifson JD, Feinberg MB, Reyes GR, McGrath MS, Stein BS, Engleman EG. AIDS retrovirus induced cytopathology: giant cell formation and involvement of CD4 antigen. Science 1986;**232**:1123-1127.
14. Lasky LA, Nakamura G, Smith DH, Fennie C, Shimasaki C, Patzer E, Berman P, Gregory T, Capon DJ. Delineation of a region of the human immunodeficiency virus type 1 gp120 glycoprotein critical for interaction with the CD4 receptor. Cell 1987;**50**:975-85.
15. Ng VL, Chiang CS, Debouk C, McGrath MS, Grove TH, Mills J. Reliable confirmation of antibodies to HIV-1 with an enzyme-linked immunoassay using recombinant antigens derived from the HIV-1 gag, pol, and env genes. J Clin Microbiol 1989;**27**:977-982.
16. Hildreth JE, Orenstas RJ. Involvement of a leukocyte adhesion receptor (LFA-1) in HIV-induced syncytium formation. Science 1989;**244**:1075-8.
17. Rostad SW, Sumi SW, Shaw C-M, Olson K, McDougall JK. HIV infection in brains with AIDS-related encephalopathy. AIDS Res and Human Retroviruses 1987;**3**:363-373.
18. Deen KC, McDougal JS, Inacker R, Folena-Wasserman G, Arthos J, Rosenberg J, Maddon PJ, Axel R, Sweet R. A soluble form of CD4 (T4) protein inhibits AIDS virus infection. Nature 1988;**331**:82-84.

19. Kahn JO, Allan JD, Hodges TL, et al. The safety and pharmacokinetics of recombinant soluble CD4 (rCD4) in subjects with the acquired immunodeficiency syndrome (AIDS) and AIDS-related complex. A phase I study. Ann Intern Med 1990;**112**:254-61.
20. Capon DJ, Chamow SM, Mordenti J, et al. Designing CD4 immunoadhesions for AIDS therapy. Nature 1989;**337**:525-31.
21. Berger EA, Clouse KA, Chaudhary VK, Chakrabarti S, Fitzgerald DJ, Pastan I, Moss B. CD4-Pseudomonas exotoxin hybrid protein blocks the spread of HIV infection in vitro and is active against cells expressing the envelope glycoproteins from diverse primate immunodeficiency viruses. Proc Natl Acad Sci USA 1989;**86**:9539-43.
22. Till MA, Ghetie V, Gregory T, et al. HIV-infected cells are killed by rCD4-ricin A chain. Science 1988;**242**:1166-8.

ACKNOWLEDGEMENTS

The authors with to acknowledge the expert technical assistance of John Marshall and Jennifer Boothman of MacFarlane Burnet Centre relating to electron microscopy and Nancy McNamara and Jean Kirihara of San Francisco General Hospital relating to macrophage preparation.

## DISCUSSION

Scofield (UCLA): You said, it's puzzling, isn't it, that you can get inhibition of infected macrophage binding to CD4-positive cells with anti-gp120, but you cannot actually demonstrate the gp120 marker's presence on the surfaces of those cells.

Mills: We have not actually shown inhibition of fusion with anti-gp120 and, to some extent, that's a trivial issue here. You need large quantities of these reagents to do the fusion assays. The anti-CD4 monoclonals that you need are in the range of 10 or 20 micrograms per milliliter, and it's very difficult to get that quantity of gp120 monoclonal antibodies from the individuals who own them.

Scofield: Well, I guess the reason I asked the question was, it may not be trivial. As you know, we study a system that has some similar features to HIV and gp120/CD4-mediated fusion: Sperm fusion to target cells via CD4 as a target ligand, with a CD4-like marker on the sperm. All we know about that is that we can inhibit the interaction between sperm and DR with anti-CD4, but we cannot, in every case demonstrate the presence of CD4 markers on the sperm's surface directly, and permeabilizing the sperm actually helps a little bit, but not much. It's a curious thing, so I wonder what you. . .

Fuller (University of Michigan): I work on herpes simplex virus and one approach we're taking to try to understand what causes syncytia there is to express the obvious glycoproteins that may be involved in cells by themselves, or in combinations. And I know there's some problem with getting gp120 or gp41 expressed constitutively or inducibly in cells, but could you comment on whether that approach is being used and what's known about whether those proteins expressed alone can affect cell fusion, or if there's anything being done in that area?

Mills: I don't think that's been done. We are not taking that approach and the reason for that is that I'm a "monomaniac." We are interested in macrophages and there are no constitutive expression systems in macrophages, as they don't replicate in culture. There are transient expression systems, but they have been difficult to work with, and the proportion of cells transfected is low.

Fuller: So, you don't know of anything being done . . .

Mills: I'm not sure, but I don't think anyone has directly done that experiment. They've certainly done the experiment, as we have, looking at CD4-positive and CD4-negative target cells, and it's very clear that you don't get fusion with CD4-negative cells.

McDougal (Centers for Disease Control): In response to that question, HIV envelope glycoproteins have been expressed transiently or stably in a variety of cell lines, and do effect fusion. The other question I had was, if you know how efficient the fusion process is, in other words, if you take a hundred cells that are infected, or p24 antigen-positive by whatever assay you're using, do you have an idea how many of those will actually enter into fusion with CD4?

Mills: I'm trying to do those experiments because I wanted to try and use that as a probe for what proportion of the cells were expressing gp120. Or to take it one step back from that, what proportion of the cells were infected. The difficulty is that fusion takes place over a couple of days. VB cells are productively infected and they also replicate, so it's very hard to go from numbers of giant cells to what proportion of the macrophages are infected. I will say that in a system in which we use between $10^4$ and $10^5$ macrophages, you get a hundred or two hundred giant cells in the culture. So, if you use that number and presume that each macrophage is only combining with VB cells, which is not a safe assumption because we can get macrophage-macrophage fusion as well, that would suggest that it is roughly between 0.1 and 1% of the cells.

Lubin (Children's Hospital, Oakland): John, I wanted to ask, if you activate the macrophages, do you enhance the expression of gp120, or have you tried that?

Mills: We haven't done that. It's not easy to quantitate gp120 expression in the fusion assay. I think if we can get good staining of the gp120 using the fluorescein-conjugated CD4, then I think we'll be in a position to actually assess expression in a more direct fashion.

Lubin: I was thinking of combinations. If you try to draw an analogy to the in vivo situation. If you secondarily infected the macrophages, or put them in an environment of lymphokines of some sort, will you enhance expression by remodeling the membrane?

Mills: Well, we have a little data on that with respect to p24. It's almost at the level of an anecdote, but with a couple of patients we have had an opportunity to follow them prospectively and assess p24 antigen expression in macrophages infected in vivo. In two cases, we had a very clear increase in antigen expression in association with a well-defined intercurrent infection. We have looked at some of the CSFs (colony stimulating factors) and we've looked at alpha-interferon. Those inhibit, or have no effect, in the case of the CSFs, so we have not seen the CSF affect that Howard Gendelman has reported. There's no increase. I had a chance to talk with Tom Matthews who's been working in Dani Bolognesi's lab and he also made the same observation. He hasn't seen any infective CSF, and gamma-interferon inhibits as well. That's been a pretty consistent observation, I think.

Ho (UCLA): John, there were a couple of reports in Montreal that suggest that infected monocyte macrophages serve as very poor targets for ADCC, or cell-mediated immunity. Do you have an explanation for that, given your findings?

Mills: It may be that the antigen expression is less than it is in lymphoid cells. I think if we can get the soluble CD4 cytofluorographic system to work, we'd be in a position to assess that. I don't think anybody's claiming that there's very much p24 expressed, although p24, at least in some systems, has been a target for ADCC, and certainly for neutralizing antibodies. So again, using a couple of different monoclonals, we have not found surface expression of p24, but it may just be that's it's insensitive and we haven't done a radioimmunoprecipitation, which might be a more sensitive way of picking that up.

Banapour (UC Davis): Do you know the level of MHC class II expression on these macrophages, and do you expect that may give you some background problem when you use your fluoresceinated CD4 on these macrophages?

Mills: If I recall correctly, and I'm speaking out of school, because I don't remember the data that well, there was not a marked change in HLA DR, DP, or DQ expression.

Düzgünes (UCSF): Any other questions or comments? I wondered whether there is anything known about why there are so many viral particles inside vacuoles in macrophages, as Gendelman has reported.

Mills: Well, our electron micrographs don't entirely fit with that. One of the reasons for doing these experiments is that you would predict from Gendelman's electron micrographs that there shouldn't be any viral proteins on the surface, but our electron micrographs, and, of course, our fusion data, clearly indicate that some gp120 is on the cell surface.

Pauza (Salk Institute): There's an old literature on that, in lentivirus infections of accumulation in intracytoplasmic vacuoles, and the old literature always said this was your standard average interferon effect. It was not specific.

Mills: Virus-generated interferon, you mean?

Pauza: No. That if you added interferon to the culture, so you're always culturing out of blood, I think those were mostly ovine lentivirus infections, where you're culturing cord blood by and large, doing a blood culture and then infecting.

Mills: You're suggesting that the vacuole formation is a function of interferon.

Pauza: Interferon treatment of those lentiviral-infected macrophages...

Mills: Induces the vacuole.

# MECHANISM OF HIV-1 ENTRY INTO CD4+ T CELLS

Barry S. Stein and Edgar G. Engleman

Stanford Blood Center
Stanford University School of Medicine
Palo Alto, CA

## Introduction

One hallmark of enveloped RNA viruses is that they are inherently fusogenic with membranes of their respective target cells. This property plays an obligate role in viral entry and is triggered and catalyzed by specific virally encoded envelope glycoproteins (1). After receptor binding by such envelope components, entry of viral genomic information into the cytosol of susceptible target cells is known to occur by two distinct mechanisms: either by direct fusion of the virus envelope with the plasma membrane in a pH-independent fashion (2), or by rapid internalization of virus via receptor-mediated endocytosis into acidic vesicles where viral envelope glycoproteins undergo requisite low pH-dependent conformational changes that facilitate virus envelope fusion with endosomal membranes (3,4). Viral entry, per se, can be strictly defined as the delivery of genomic viral RNA into the cytoplasmic compartment of host cells consequent to membrane fusion. Therefore, the simple incorporation of a virion into an endocytic vesicle, in and by itself, does not constitute viral entry in the true sense, unless the virus envelope subsequently fuses with the endosomal membrane. Endocytosed virus particles which fail to undergo intravesicular membrane fusion are likely destined for lysosomal degradation.

## The Role of CD4 in HIV-1 Entry

CD4 has been definitively established to be the cell surface receptor responsible for HIV-1 target cell tropism (5,6). HIV-1 is known to infect CD4+ T cells (7), as well as a variety of other CD4-bearing cell types, principally antigen presenting cells such as monocytes and macrophages (8-10), Langerhans cells (11), and human peripheral blood dendritic cells (12). A specific interaction between CD4 on T cells and nonpolymorphic determinants of MHC class II on antigen presenting cells has been purported to be essential to helper T cell function (13-15). However, the MHC-binding and HIV-binding domains of CD4 are partly separable (16). Activation of CD4+ T cells with appropriate antigen-bearing target cells or phorbol esters results in phosphorylation and subsequent internalization of CD4 (17,18). Similarly, binding of HIV-1 to CD4+ T cell clones has been shown to induce rapid and sustained CD4 phosphorylation which appears to involve protein kinase C (19). However, endocytosis of CD4 has been demonstrated not to be requisite for infection of CD4-bearing cells by HIV-1 (20). Studies from our laboratory provided conclusive evidence that HIV-1 enters a CD4+ T lymphoblastoid cell line (VB) via direct fusion with the plasma membrane in a pH-independent manner (21). Subsequent work with VSV (HIV-1) pseudotypes confirmed our finding that HIV-1 entry into CD4+ T cells is mediated by a pH-independent mechanism (22). Nonetheless, CD4 receptor- mediated endocytosis of HIV-1 particles has been observed in another human T lymphoid line (CEM) (23), despite the fact that we could not demonstrate any evidence for an operant endocytic HIV-1 entry pathway in VB cells (21). This may be explained on the basis that CD4 internalization is differentially regulated in different CD4+ T cell lines. More specifically, CEM cells may constitutively internalize CD4, whereas internalization of CD4 in VB cells appears to require a triggering stimulus such as phorbol ester (18). Although, HIV-1 binding has been shown to induce phosphorylation of CD4 in T cell clones (19), the overwhelming consensus of published findings indicate that HIV-1 entry does not require internalization of the CD4 molecule (20,24,25).

## HIV-1 Spike Envelope Glycoprotein Determinants and Viral Entry

gp120 and gp41 are derived from the endoproteolytic cleavage products of the gp160 kDa HIV-1 envelope polyprotein precursor (26,27,28) and are associated solely by noncovalent molecular interactions (29). Endoproteolytic cleavage of gp160 exposes the fusogenic domain at the amino terminus of gp41, which is obligate for viral infectivity (30). The native form of the gp41 HIV-1 envelope substituent has been shown to exist as a tetramer in viral and cellular membranes (31). The entire HIV-1 spike glycoprotein is therefore postulated to be an oligomer composed of four gp41 and four gp120 subunits.

The gp120 envelope moiety is responsible for virion specific binding to cell surface CD4 (5,6,32). This interaction is thought to induce the gp41 transmembrane envelope component to initiate fusion between the virus envelope and the plasma membrane of susceptible target cells, resulting in the delivery of HIV-1 RNA into the cytosol (1,30,33). Similar, if not identical, interactions between CD4 on uninfected T cells and gp120 expressed on the surface of virally infected cells lead to syncytia formation via cell-to-cell fusion (34,35). The dissociation constant ($K_D$) of the CD4/gp120 binding interaction has been estimated to be approximately $10^{-9}$M (36), with specific binding domains localizing to the carboxy terminus of gp120 (36,37) and the amino terminus of CD4 (38,39).

Little is known about the precise cascade of molecular events which occur between the initial binding of HIV-1 to cell surface CD4 and the culmination of gp41-mediated virus-to-cell membrane fusion. It appears, nonetheless, that the HIV-1 envelope spike glycoprotein complex (gp120/gp41), unlike the hemagglutinin envelope glycoprotein of influenza virus ($HA_1/HA_2$) (1,40), does not require a low pH-dependent conformational change in its tertiary structure in order to be rendered fusogenic (21,22). Alternatively, it has not been established whether binding of CD4 to gp120 confers fusogenicity to the gp120/gp41 envelope spike complex by triggering a pH-independent conformational rearrangement of hydrobic domains, or whether the native envelope spike complex is fusogenic irrespective of CD4 binding. The observation that cells expressing HIV-1 envelope glycoproteins preferentially form syncytia with CD4-positive, but not with CD4-negative cells (34,35) favors the former hypothesis. However, the possibility cannot be excluded that the binding of CD4 to gp120 functions solely to bring apposing membranes in close enough proximity for fusion to proceed spontaneously.

## Mechanism of gp41-Mediated Virus-to-Cell Membrane Fusion

The molecular mechanisms by which enveloped viruses achieve membrane fusion are poorly understood, except for the fact that viral envelope spike glycoproteins play an indispensible role in the fusion process (1). In particular, studies utilizing a recombinant vaccinia virus containing the entire HIV-1 envelope gene coding region, demonstrate that cell-surface expression of HIV-1 envelope substituents, in the absence of any other HIV-1 structural or regulatory protein, is sufficient to induce CD4-dependent cell fusion (35). Studies employing site-directed mutagenesis suggest that endoproteolytic cleavage of the HIV-1 envelope protein precursor is necessary for full expression of viral infectivity (30), which is consistent with what has been observed for other enveloped mammalian viruses such as Sendai (41-43) and influenza (44,45). Cleavage of various viral envelope protein precursors has been shown to result in the exposure of unique, yet highly conserved, hydrophobic domains which are postulated to participate directly in membrane fusion (1,30). In the case of HIV-1, the hydrophobic domain resides at the N-terminus of gp41 (29,30,37). If we accept the hypothesis that the N-terminus of gp41 is directly involved in the process of membrane fusion, then we are still left with the problem of elucidating the precise mechanism whereby the exposed hydrophobic domain initiates the fusion process.

Although the actual molecular mechanism by which viral membrane fusion occurs is not well understood for any virus, it is reasonable to assume that the enveloped viruses have exploited to their advantage fusion reactions which closely resemble those which take place between intracellular vesicles during normal cellular membrane trafficking. Two molecular mechanisms can be proposed to explain how viral envelope proteins mediate membrane fusion. Firstly, fusion could be evoked simply via the passive insertion of viral envelope hydrophobic domains (e.g., the N-terminus of gp41) into apposed lipid bilayers. Alternatively, fusion may first require specific binding of viral hydrophobic domains to putative receptor(s). As precedence for the latter hypothesis, it has been postulated that translocation of nascent polypeptides across the endoplasmic reticulum is initiated by recognition of their hydrophobic N-terminal signal peptides by the hydrophobic domain of a translocator protein complex present on the cytosolic side of the endoplasmic reticulum (46,47). It is not unreasonable to speculate that an analogous translocator complex may also exist at the cell surface, which is exploited by viral spike glycoproteins to mediate membrane fusion. Nevertheless, there are numerous examples where viral envelope glycoproteins have been reported to fuse with lipid bilayers devoid of protein (48-54). Therefore, the question as to whether virus induced membrane fusion proceeds via passive insertion of envelope hydrophobic domains into target membranes, or via a specific binding interaction between envelope hydrophobic domains and a putative protein receptor on the surface of target membranes remains unsettled.

The fusion of amphiphilic membrane bilayers is strongly influenced by interbilayer interactions, which include attractive van der Waals and hydrophobic forces as well as repulsive hydration and electrostatic forces (55-63). Studies of artificial lipid membranes have established that spontaneous fusion between two approaching bilayers occurs at a low frequency (1). The low fusion frequency was believed to be a consequence of the magnitude of repulsive hydration forces (1,64). However, a recent report indicates that membrane bilayers do not have to overcome a repulsive force barrier such as hydration, before fusion can occur (66); rather, once bilayers are apposed within approximately 1 nm of each other, local deformations and molecular rearrangements bypass repulsive forces. In addition, the major force facilitating membrane fusion appears to be the hydrophobic attraction of internal hydrocarbon chains between two apposing bilayers, suggesting that attractive forces at external membrane surfaces only suffice to mediate bilayer-to-bilayer adhesion (65). Based upon these observations, a general model for enveloped virus-to-cell membrane fusion can be constructed. An infectious virion must first achieve intimate contact with the cell surface of its respective target cell to initiate bilayer adhesion. In the case of HIV-1, the high affinity binding interaction between CD4 and gp120 facilitates this process by overcoming repulsive hydration and electrostatic forces. Exposure of hydrophobic interiors of the two apposing membranes then takes place secondary to highly localized molecular rearrangements, and culminates in membrane fusion. The N-terminal hydrophobic domain of gp41 apparently potentiates and is paramount to the interaction of hydrophobic interiors between viral and cellular membrane, however, the manner in which this is initiated still remains undefined.

One question that deserves consideration is whether gp41 in and by itself is capable of initiating membrane fusion in the absence of gp120. It is not unreasonable to postulate that naked gp41 oligomeric complexes (31) devoid of gp120 are generated on the surface of virions and productively infected cells as a consequence of molecular shedding of gp120. However, HIV-1 infected cells have only been noted to form syncytia with cells that express surface CD4 (35), which argues against the idea that gp120-CD4 binding is dispensable to the fusion process. This is further supported by the general observation that the susceptibility of a given cell to infection by HIV-1 is dependent upon the expression of CD4 (5-7), with the apparent exception of certain human tumor cell lines in which an alternative receptor may be involved (66,67). Therefore, the fact that spontaneous gp41-mediated membrane fusion appears to be thermodynamically unfavored in the absence of gp120-CD4 binding strongly diminishes, albeit does not eliminate, the possibility that such fusion is biologically relevant. The strong binding interaction between CD4 and gp120 likely suffices to overcome force barriers limiting membrane-to-membrane adhesion, so that fusion can proceed spontaneously. In the more general sense, it is not surprising that the enveloped viruses have evolved in such a manner that their spike glycoproteins contain specific domains which serve as high affinity ligands for specific receptors on respective target cells (1); thus, establishing intimate viral-to-cell membrane adhesion reactions permissive to spontaneous fusion.

## CD4-Independent HIV-1 Entry

CD4-independent HIV-1 infection has been reported in certain glioma and rhabdomysosarcoma cell lines (66). In particular, HIV-1 infection could not be blocked in these cell lines by either soluble CD4 or anti-CD4 monoclonal antibody. These observations lead to the suggestion that there may be another receptor for HIV-1 other than CD4, or that HIV-1 infection occurs by gp41-mediated membrane fusion in the absence of a specific gp120-CD4 receptor interaction. Further studies are needed to more precisely define the molecular basis for any putative non-CD4 mediated HIV-1 entry pathways. The most critical question, however, remains whether non-CD4 mediated HIV-1 entry pathways described in in vitro systems are salient to the immunopathogenesis of AIDS in infected individuals.

## Mechanism of Entry of HIV-1 Via Antibody Enhancement

There is also evidence that the FcR III receptor on human macrophages, and possibly another Fc receptor on human CD4+ lymphocytes, mediate antibody-dependent enhancement of HIV-1 infectivity in vitro (68). This enhancement is postulated to be independent of CD4, inasmuch as there is no significant blocking encountered with high concentrations of anti-Leu 3a monoclonal antibody or soluble CD4. Nonetheless, viral envelope fusion with either the plasmalemma or endosomal membrane must occur for HIV-1 infection to proceed, regardless of whether viral entry took place in the absence or presence of antibody enhancement. Thus, it is important to construct a model which includes a membrane fusion event during the course of Fc receptor-mediated enhancement of HIV-1 infection. In principal, one could envision that anti-HIV-1 antibody complexes containing bound HIV-1 virions at Fab domains initially attach to cell surface Fc receptors. The Fc receptor-antibody-virion complexes then undergo receptor mediated endocytosis (69). Once in the endosomes, rapid acidification occurs, and as a result antibody-virion complexes likely dissociate from Fc receptors. It can also be postulated that antibody-virion complexes themselves are subject to dissociation secondary to acidic pH. Under these circumstances the resultant free virions and free antibody-virion complexes may undergo fusion with endosomal membranes to initiate infection. If internalized CD4 is also present in the same endosome, then classic CD4-mediated viral entry can explain viral infection. The observation that anti-Leu 3a monoclonal antibody and soluble CD4 do not prevent enhancement argues against this possibility (68). However, anti-Leu 3a monoclonal antibody could conceivably dissociate from CD4 during endosomal acidification, abrogating its blocking effects on CD4-mediated HIV-1 infection. In addition, the achievable concentration of soluble CD4 present in the microenvironment of an endosome may be insufficient to completely block entry of HIV-1 virions via the CD4 receptor. On the other hand, if anti-Leu 3a monoclonal antibody and soluble CD4 do efficiently block CD4-mediated viral entry within endosomal compartments, or the endosomal compartments in fact do not contain any CD4, then one could postulate that there is an alternative HIV-1 receptor involved or that gp120-independent gp41-mediated fusion takes place.

Role of T Cell Activation in HIV-1 Entry

Earlier reports had indicated that among T cells already infected with HIV-1, subsequent T cell activation greatly enhances viral gene expression leading to the production of infectious virions (70,71). Although these observations have added to a growing body of evidence that T cell activation is an important determinant for HIV-1 gene expression, they do not address the possible role T cell activation plays in viral nucleocapsid entry into the cytoplasm. Initial studies undertaken in our laboratory indicated that short-term activation of unstimulated peripheral blood CD4+ T cells clearly facilitates the generation of unintegrated HIV-1 DNA duplex forms upon exposure to infectious virions (72). More specifically, unintegrated HIV-1 DNA was not detected by Southern analysis in unstimulated CD4+ T cells exposed to HIV-1 (M.O.I. ~5) for periods as long as 48 hours. However, unintegrated HIV-1 DNA was easily detected in CD4+ T cells prestimulated with either PHA or OKT3 monoclonal antibody for 1 hour and subsequently exposed to virus for as short a time interval as 2 hours. These data suggest that the ultimate reverse transcription of the HIV-1 RNA genome into DNA is dependent upon the cascade of events induced by early T cell activation. In particular, viral-to-cell membrane fusion, which results in entry of viral nucleocapsid cores into the cytosol proper, may require an antecedent T cell activation stimulus in order to proceed spontaneously. Nonetheless, activation may also be indispensable to post nucleocapsid entry events such as core unraveling and/or reverse transcriptase-directed DNA strand synthesis.

Further studies provided evidence that unstimulated CD4+ T cells are relatively, if not completely, resistant to viral-to-cell membrane fusion (72). Unstimulated CD4+ T cells were co-cultured with equal numbers of HXB/gpt cells (a T cell line which constitutively expresses high levels of surface gp120/gp41 viral spike envelope glycoproteins) and were monitored for the appearance of multinucleated giant cells. No evidence of cell-to-cell fusion was observed for at least 2 hours, and thereafter, small giant cells with relatively few nuclei began to form. In contrast, CD4+ T cells prestimulated with either PHA or OKT3 monoclonal antibody for periods as short as 1 hour began to form large giant cells within 2 hours of exposure to HXB/gpt cells. Thus, early T cell activation signals enhance cell-to-cell fusion and giant cell formation. The delayed appearance of multinucleated giant cells in co-cultures containing unstimulated CD4+ T cells and HXB/gpt cells is most likely due to allogeneic stimulation. We interpret these results as indicating that resting CD4+ T cells are relatively resistant to viral fusion, and therefore are not permissive to nucleocapsid entry into the cytosol. It is salient to note that cell surface CD4 antigen density does not vary significantly after delivery of activation stimuli. In addition, soluble gp120 binds to unstimulated and activated CD4+ T cells with equal affinity.

To test the hypothesis that early T cell activation stimuli confer or enhance the fusogenic potential of the plasmalemma of resting cells with the envelope of HIV-1, we employed a number of known inhibitors of T cell activation and assayed their ability to prevent fusion between unstimulated CD4+ T cells and the HXB/gpt cell line. The results indicated that while inhibitors of relatively late stages of T cell activation (i.e., cyclosporin and mitomycin C) had no effect on the formation of syncytia, inhibitors of protein kinase C activity (i.e., H7 and cholera toxin) completely blocked the formation of syncytia in mixtures of fresh unstimulated CD4+ T cells and HXB/gpt cells in a dose-dependent and reversible fashion (73). The inhibition induced by cholera toxin could be overcome by the addition of 1-oleoyl-2-acetylglycerol, a cell penetrating analog of diacylglycerol, but not by ionomycin, a calcium ionophore. This strongly suggests that protein kinase C activation is required to render otherwise resting CD4+ T cells fusogenic (74). However, both H7 and cholera toxin failed to inhibit the ability of T cells activated with either PHA or OKT3 monoclonal antibody for 72 hours to form syncytia with HXB/gpt cells. This observation implies that the fusogenic potential of CD4+ T cells for HIV-1 envelope bearing membranes is dependent upon PKC activity only during the early stages of activation.

Taken together, our findings provide strong evidence that T cell activation significantly enhances, and is likely requisite for HIV-1 entry into CD4+ T cells. Derived from this thesis is the idea that resting CD4+ T cells are resistant to HIV-1 infection. In the interpretation of our data leading to this conclusion we had made the assumption that HIV-1 envelope mediated cell-to-cell syncytia formation is mechanistically analogous to virus-to-cell membrane fusion. However, one must be cautious in that one fundamental difference does distinguish the two fusion processes, in that virion fusion only involves small stretches of membrane less than 100 nM in length whereas cell-to-cell fusion occurs across a distance of many microns. This may help to explain why virions can fuse to cell membranes at 4°C

(21), whereas cell-to-cell fusion is completely abrogated at this temperature. It would therefore be logical to assume that cell-to-cell fusion requires more energy to proceed spontaneously than virus-to-cell fusion secondary to a greater requirement for membrane deformability. However, except for this thermodynamic factor, the two fusion processes are likely otherwise identical.

The precise molecular basis which would explain why resting CD4+ T cells are inherently resistant to HIV-1 envelope-mediated membrane fusion remains elusive. Nonetheless, our findings provide compelling evidence that PKC-mediated phosphorylation reactions that are induced during early stages of T cell activation play an essential role in conferring fusogenic properties to resting CD4+ T cells. De novo protein synthesis is apparently not required for resting cells to be rendered fusogenic with HXB/gpt cells upon allogeneic stimulation in co-culture, inasmuch as cycloheximide exhibits no inhibitory effect on this process. Therefore it is likely that PKC-mediated phosphorylation reactions which take place during early T cell activation serve to induce spontaneous membrane fusogenicity by rapidly augmenting membrane deformability. Alternatively, if there is a specific receptor for gp41 which mediates membrane fusion, PKC-mediated phosphorylation reactions may be required to translocate the putative receptor from the cytosol to the plasma membrane, or to induce molecular conformational changes requisite for gp41 recognition.

The observation that resting CD4+ T cells do not appear to generate HIV-1 DNA forms upon exposure to virus (72), confirms our hypothesis that resting cells are resistant to HIV-1 entry. However, we cannot exclude the possibility that Southern analysis lacks the sensitivity to measure small copy numbers of the viral genome. PCR analysis would definitively increase the sensitivity of detection. However, this technology is limited in that any contaminating HIV-1 DNA present in viral inoculums could then serve as a false positive template, generating signal even in cells extensively washed free of virions. Furthermore, virtually any population of T cells derived from peripheral blood contains a small percentage of in vivo activated cells, and such cells would be expected to be permissive to HIV-1 entry, possibly yielding signal detectable by PCR, but not by Southern analysis. Despite the fact that our studies indicate that HIV-1 does not generate DNA forms in resting CD4+ T cells secondary to restricted viral entry, we do not have any evidence as to whether the events of early T cell activation are directly required for nucleocapsid unraveling and/or reverse transcription of the viral RNA genome itself. Perhaps future studies employing in vitro reverse transcription assays will be able to better address this question.

As a final statement, with regard to entry of HIV-1 into resting CD4+ T cells, it can be said that PKC-mediated phosphorylation reactions which are induced during early T cell activation, confer or at least drastically augment virus-to-cell and cell-to-cell HIV-1 mediated fusion reactions. It is probably more appropriate to conclude that spontaneous fusion between membranes of resting CD4+ T cells and the HIV-1 envelope is strongly unfavored thermodynamically. This analysis would then imply that HIV-1 entry into resting CD4+ T cells is not absolutely restricted and that infection of such cells may in fact occur, albeit rarely. In the final analysis, one is left with the question as to the clinical relevance of this experimental observation. It has been reported that the reservoir for HIV-1 in human peripheral blood is the CD4+ T cell (75). The frequency of CD4+ T cells that contain HIV-1 DNA is estimated to be at least 1/100 in patients with AIDS. In addition, titers of HIV-1 in the plasma of AIDS patients have been found to be higher than previously estimated; at approximately 1-2 x $10^3$ tissue-culture-infective doses per ml of plasma (76,77). Despite the viremia, the fact that only 1/100 cells contain HIV-1 DNA, and 1/1000 cells express HIV-1 proteins (75), is consistent with the notion that resting CD4+ T cells are resistant to viral entry. If resting cells were not resistant, one would principally expect to observe efficient infection of virtually all peripheral blood CD4+ T cells in AIDS patients as a consequence of the high virus titer present in the plasma.

## Summary

Although the mechanism responsible for HIV-1 entry into susceptible CD4+ T cells is incompletely understood, a number of key components are now known. For example, the tropism of HIV-1 for cells expressing the CD4 membrane glycoprotein reflects the use of this protein as a specific viral receptor to which the HIV-1 gp120 envelope protein binds with high affinity. This binding apparently results in the exposure of hydrophobic domains of the gp41 transmembrane protein to apposing plasma membrane components, resulting in the fusion of viral and plasma membranes to one another which, in turn, releases HIV-1 RNA into the cytosol. This fusion event, which is requisite for viral entry as well as HIV-1 associated syncytia formation, occurs in a pH-independent fashion, but requires antecedent T cell activation. In the absence of T cell stimuli, resting CD4+ cells are resistant to HIV-1 entry, which may explain the observation that at any given time the vast majority of CD4+ T cells in HIV-1 seropositive patients are not infected despite the presence of relatively large quantities of free virus in the blood of such patients. The mechanism of HIV-1 entry into other CD4+ cell types, such as macrophages and dendritic cells, remains to be determined.

References


1. J. White, M. Kielian, and A. Helenius, Membrane fusion proteins of enveloped animal viruses, Quart. Rev. Biophys. 16:151 (1983).
2. K. Apostolov and J. D. Almeida, Interaction of Sendai virus (HV) with human erythrocytes: a morphological study of hemolysis and cell fusion, J. Gen. Virol. 15:227 (1972).
3. A. Helenius, J. Kartenbeck, K. Simons, and E. Fries, On entry of Semliki forest virus into BHK-21 cells, J. Cell Biol. 84:404 (1980).
4. K. S. Matlin, H. Reggio, A. Helenius, and K. Simons, Infectious entry pathway of influenza virus in a canine kidney cell line, J. Cell Biol. 91:601 (1981).
5. A. G. Dalgleish, P. C. L. Beverley, P. R. Clapham, D. H. Crawford, M. F. Greaves, and R. A. Weiss, The CD4 (T4) antigen is an essential component of the receptor for the AIDS retrovirus, Nature 312:763 (1984).
6. D. Klatzmann, E. Champagne, S. Chamaret, J. Gruest, D. Guetard, T. Hercend, J.-C. Gluckman, and L. Montagnier, T-lymphocyte T4 molecule behaves as the receptor for human retrovirus LAV, Nature 312:767 (1984).
7. D. Klatzmann, F. Barre-Sinoussi, M. T. Nugeyre, C. Dauget, E. Vilmer, C. Griscelli, F. Brun-Vezinet, C. Rouzioux, J.-C. Chermann, and L. Montagnier, Selective tropism of lymphadenopathy associated virus (LAV) for helper-inducer T lymphocytes, Science 225:59 (1984).
8. S. Gartner, P. Markovits, D. M. Markovitz, M. H. Kaplan, R. C. Gallo, and M. Popovic, The role of mononuclear phagocytes in HTLV-III/LAV infection, Science 233:215 (1986).
9. D. D. Ho, T. R. Rota, and M. S. Hirsch, Infection of monocyte/macrophages by human T lymphotropic virus type III, J. Clin. Invest. 77:1712 (1986).
10. J. K. A. Nicholson, G. D. Cross, C. S. Callaway, and J. S. McDougal, In vitro infection of human monocytes with human T lymphotropic virus type III/lymphadenopathy-associated virus (HTLV-III/LAV), J. Immunol. 137:323 (1986).
11. E. Tschachler, V. Groh, M. Popovic, D. L. Mann, K. Konrad, B. Safai, L. Eron, F. D. Veronese, K. Wolff, and G. Stingl, Epidermal Langerhans cells - a target for HTLV-III/LAV infection, J. Invest. Dermatol. 88:233 (1987).
12. S. Patterson and S. C. Knight, Susceptibility of human peripheral blood dendritic cells to infection by human immunodeficiency virus, J. Gen. Virol. 68:1177 (1987).
13. E. G. Engleman, C. Benike, C. Grumet, and R. L. Evans, Activation of human T lymphocyte subsets: helper and suppressor/cytotoxic T cells recognize and respond to distinct histocompatibility antigens, J. Immunol. 127:2124 (1981).

14. S. L. Swain, T cell subsets and the recognition of MHC class, Immunol. Rev. 74:129 (1983).
15. D. Gay, P. Maddon, R. Sekaly, M. A. Talle, M. Godfrey, E. Long, G. Goldstein, L. Chess, R. Axel, J. Kappler, and P. Marrack, Functional interaction between human T-cell protein CD4 and the major histocompatibility complex HLA-DR antigen, Nature 328:626 (1987).
16. D. Lamarre, A. Ashkenazi, S. Fleury, D. H. Smith, R.-P. Sekaly, and D. J. Capon, The MHC-binding and gp120-binding functions of CD4 are separable, Science 245:743 (1989).
17. R. B. Acres, P. J. Conlon, D. Y. Mochizuki, and B. Gallis, Rapid phosphorylation and modulation of the T4 antigen induced by phorbol myristate acetate or antigen, J. Biol. Chem. 261:16210 (1986).
18. J. A. Hoxie, D. M. Matthews, K. J. Callahan, D. L. Cassel, and R. A. Cooper, Transient modulation and internalization of T4 antigen induced by phorbol esters. J. Immunol. 137:1194 (1986).
19. A. P. Fields, D. P. Bednarik, A. Hess, and W. S. May, Human immunodeficiency virus induces phosphorylation of its cell surface receptor, Nature 333:278 (1988).
20. P. J. Maddon, J. S. McDougal, P. R. Clapham, A. G. Dalgleish, S. Jamal, R. A. Weiss, and R. Axel, HIV infection does not require endocytosis of its receptor, CD4, Cell 54:865 (1988).
21. B. S. Stein, S. D. Gowda, J. D. Lifson, R. C. Penhallow, K. G. Bensch, and E. G. Engleman, pH-independent HIV entry into CD4-positive T cells via virus envelope fusion to the plasma membrane, Cell 49:659 (1987).
22. M. O. McClure, M. Marsh, and R. A. Weiss, Human immunodeficiency virus infection of CD4-bearing cells occurs by a pH-independent mechanism, EMBO J. 7:513 (1988).
23. C. D. Pauza and T. M. Price, Human immunodeficiency virus infection of T cells and monocytes proceeds via receptor-mediated endocytosis, J. Cell Biol. 107:959 (1988).
24. J. A. Hoxie, J. L. Rackowski, B. S. Haggerty, and G. N. Gaulton, T4 endocytosis and phosphorylation induced by phorbol esters but not by mitogen or HIV infection, J. Immunol. 140:786 (1988).
25. P. Bedinger, A. Moriarty, R. C. von Borstel II, N. J. Donovan, K. S. Steimer, and D. R. Littman, Internalization of the human immunodeficiency virus does not require the cytoplasmic domain of CD4, Nature 334:162 (1988).
26. J. S. Allan, J. E. Coligan, F. Barin, M. F. McLane, J. G. Sodroski, C. A. Rosen, W. A. Haseltine, T. H. Lee, and M. Essex, Major glycoprotein antigens that induce antibodies in AIDS patients are encoded by HTLV-III, Science 228:1091 (1985).
27. F. D. Veronese, A. L. DeVico, T. D. Copeland, S. Oroszlan, R. C. Gallo, and M. G. Sarngadharan, Characterization of

gp41 as the transmembrane protein coded by the HTLV-III/LAV envelope gene, Science 229:1402 (1985).

28. R. L. Willey, J. S. Bonifacino, B. J. Potts, M. A. Martin, and R. D. Klausner, Biosynthesis, cleavage, and degradation of the human immunodeficiency virus 1 envelope glycoprotein gp160, Proc. Natl. Acad. Sci. USA 85:9580 (1988).
29. S. Modrow, B. H. Hahn, G. M. Shaw, R. C. Gallo, F. Wong-Staal, and H. Wolf, Computer-assisted analysis of envelope protein sequences of seven human immunodeficiency virus isolates: prediction of antigenic epitopes in conserved and variable regions, J. Virol. 61:570 (1987).
30. J. M. McCune, L. B. Rabin, M. B. Feinberg, M. Lieberman, J. C. Kosek, G. R. Reyes, and I. L. Weissman, Endoproteolytic cleavage of gp160 is required for the activation of human immunodeficiency virus, Cell 53:55 (1988).
31. A. Pinter, W. J. Honnen, S. A. Tilley, C. Bona, H. Zaghouani, M. K. Gorney, and S. Zolla-Pazner, Oligomeric structure of gp41, the transmembrane protein of human immunodeficiency virus type 1, J. Virol. 63:2674 (1989).
32. J. S. McDougal, M. S. Kennedy, J. M. Sligh, S. P. Cort, A. Mawle, and J. K. A. Nicholson, Binding of HTLV-III/LAV to T4+ T cells by a complex of the 110K viral protein and the T4 molecule, Science 231:382 (1986).
33. P. W. Berman, W. M. Nunes, and O. K. Haffar, Expression of membrane-associated and secreted variants of gp160 of human immunodeficiency virus type 1 in vitro and in continuous cell lines, J. Virol. 62:3135 (1988).
34. J. D. Lifson, G. R. Reyes, M. S. McGrath, B. S. Stein, and E. G. Engleman, AIDS retrovirus induced cytopathology: giant cell formation and involvement of CD4 antigen, Science 232:1123 (1986).
35. J. D. Lifson, M. B. Feinberg, G. R. Reyes, L. Rabin, B. Banapour, S. Chakrabarti, B. Moss, F. Wong-Staal, K. S. Steimer, and E. G. Engleman, Induction of CD4-dependent cell fusion by the HTLV-III/LAV envelope glycoprotein, Nature 323:725 (1986).
36. L. A. Lasky, G. Nakamura, D. H. Smith, C. Fennie, C. Shimasaki, E. Patzer, P. Berman, T. Gregory, and D. J. Capon, Delineation of a region of the human immunodeficiency virus type 1 gp120 glycoprotein critical for interaction with the CD4 receptor, Cell 50:975 (1987).
37. M. Kowalski, J. Potz, L. Basiripour, T. Dorfman, W. C. Goh, E. Terwilliger, A. Dayton, C. Rosen, W. Haseltine, and J. Sodroski, Functional regions of the envelope glycoprotein of human immunodeficiency virus type 1, Science 237:1351 (1987).
38. N. R. Landau, M. Warton, and D. R. Littman, The envelope glycoprotein of the human immunodeficiency virus binds to the immunoglobulin-like domain of CD4, Nature 334:159 (1988).

39. A. Peterson and B. Seed, Genetic analysis of monoclonal antibody and HIV binding sites on the human lymphocyte antigen CD4, Cell 54:65 (1988).
40. J. Skehel and M. Waterfield, Studies on the primary structure of the influenza virus hemagglutinin, Proc. Natl. Acad. Sci. USA 72:93 (1982).
41. M. Homma and M. Ohuchi, Trypsin action on the growth of Sendai virus in tissue culture cells, J. Virol. 12:1257 (1973).
42. A. Scheid and P. W. Choppin, Identification of biological activities of paramyxovirus glycoproteins. Activation of cell fusion, hemolysis, and infectivity by proteolytic cleavage of an inactive precursor protein of Sendai virus, Virology 57:475 (1974).
43. A. Scheid and P. W. Choppin, Protease activation mutants of Sendai virus. Activation of biological properties by specific proteases, Virology 69:265 (1976).
44. S. Lazarowitz, R. W. Compans, and P. W. Choppin, Proteolytic cleavage of the hemagglutinin polypeptide of influenza virus: function of the uncleaved polypeptide HA, Virology 68:199 (1973).
45. H.-D. Klenk, R. Rott, M. Orlich, and J. Blodorn, Activation of influenza A viruses by trypsin treatment, Virology 68:426 (1975).
46. C. Blobel, Intracellular protein topogenesis, Proc. Natl. Acad. Sci. USA 77:1496 (1980).
47. S. J. Singer, P. A. Maher, and M. P. Yaffe, On the translocation of proteins across membranes, Proc. Natl. Acad. Sci. USA 84:1015 (1987).
48. J. J. Skehel, P. Bayley, E. Brown, S. Martin, M. Waterfield, J. White, I. Wilson, and D. Wiley, Changes in the conformation of influenza virus hemagglutinin at the pH optimum of virus-mediated membrane fusion, Proc. Natl. Acad. Sci. USA 79:968 (1982).
49. T. Maeda, K. Kawasaki, and S.-I. Ohnishi, Interaction of influenza virus hemagglutinin with target lipids is a key step in virus-induced hemolysis and fusion at pH 5.2, Proc. Natl. Acad. Sci. USA 78:4133 (1981).
50. G. VanMeer, J. Davoust, and K. Simons, Parameters affecting low pH-mediated fusion of liposomes with the plasma membrane of cells infected with influenza virus, Biochemistry 24:3593 (1985).
51. A. M. Haywood, Fusion of Sendai viruses with model membranes, J. Mol. Biol. 87:625 (1974).
52. J. White and A. Helenius, pH-dependent fusion between the Semliki Forest virus membrane and liposomes, Proc. Natl. Acad. Sci. USA 77:3273 (1980).
53. J. J. Mooney, J. M. Dalrymple, C. R. Alving, and P. K. Russell, Interaction of Sindbis virus with liposomal model membranes, J. Virol. 15:225 (1975).

54. R. K. Scheule, Fusion of Sindbis virus with model membranes containing phosphatidylethanolamine: implications for protein induced membrane fusion, Biochim. Biophys. Acta 899:185 (1987).
55. L. V. Chernomordik, G. B. Melikyan, and Y. A. Chizmadzhev, Biomembrane fusion: a new concept derived from model studies using two interacting planar lipid bilayers, Biochim. Biophys. Acta 906:309 (1987).
56. R. P. Rand and V. A. Parsegian, Mimicry and mechanism in phospholipid models of membrane fusion, Ann. Rev. Physiol. 48:201 (1986).
57. D. M. LeNeveu, R. P. Rand, V. A. Parsegian, and D. Gingell, Measurement of forces between lecithin bilayers, Nature 259:601 (1976).
58. D. M. LeNeveu, R. P. Rand, V. A. Parsegian, and D. Gingell, Measurement and modification of forces between lecithin bilayers, Biophys. J. 18:209 (1977).
59. L. J. Lis, M. McAlister, N. Fuller, R. P. Rand, and V. A. Parsegian, Interactions between neutral phospholipid bilayer membranes, Biophys. J. 37:657 (1982).
60. P. Claesson, A. M. Carmona-Ribeiro, and K. J. Kurihara, Dihexadecyl phosphate monolayers - intralayer and interlayer interactions, J. Phys. Chem. 93:917 (1989).
61. J. N. Israelachvili and R. M. Pashley, The hydrophobic interaction is long range, decaying exponentially with distance, Nature 300:341 (1982).
62. R. M. Pashley, P. M. McGuiggan, B. W. Ninham, and D. F. Evans, Attractive forces between uncharged hydrophobic surfaces: direct measurements in aqueous solution, Science 229:1088 (1985).
63. P. M. Claesson, C. E. Blom, P. C. Herder, and B. W. Ninham, Interactions between water-stable hydrophobic Langmuir-Blodgett monolayers on mica, J. Colloid Interface Sci. 114:234 (1986).
64. R. Rand, Interacting phospholipid bilayers: measured forces and induced structural changes, Ann. Rev. Biophys. Bioeng. 10:277 (1981).
65. C. A. Helm, J. N. Israelachvili, and P. M. McGuiggan, Molecular mechanisms and forces involved in the adhesion and fusion of amphiphilic bilayers, Science 246:919 (1989).
66. P. R. Clapham, J. N. Weber, D. Whitby, K. McIntosh, A. G. Dalgleish, P. J. Maddon, K. C. Deen, R. W. Sweet, and R. A. Weiss, Soluble CD4 blocks the infectivity of diverse strains of HIV and SIV for T cells and monocytes but not for brain and muscle cells, Nature 337:368 (1989).
67. M. Tateno, F. Gonzalez-Scarano, and J. A. Levy, Human immunodeficiency virus can infect CD4-negative human fibroblastoid cells, Proc. Natl. Acad. Sci. USA 86:4287 (1989).

68. J. Homsy, M. Meyer, M. Tateno, S. Clarkson, and J. A. Levy, The Fc and not CD4 receptor mediates antibody enhancement of HIV infection in human cells, Science 244:1357 (1989).
69. I. Mellman and H. Plutner, Internalization and degradation of macrophage Fc receptors bound to polyvalent immune complexes, J. Cell Biol. 98:1170 (1984).
70. D. Zagury, J. Bernard, R. Leonard, R. Cheynier, M. Feldman, P. S. Sarin, and R. C. Gallo, Long-term cultures of HTLV-III-infected T cells: a model of cytopathology of T-cell depletion in AIDS, Science 231:850 (1986).
71. J. B. Margolick, D. J. Volkman, T. M. Folks, and A. S. Fauci, Amplification of HTLV-III/LAV infection by antigen-induced activation of T cells and direct suppression by virus of lymphocyte blastogenic responses, J. Immunol. 138:1719 (1987).
72. S. D. Gowda, B. S. Stein, N. Mohagheghpour, C. J. Benike, and E. G. Engleman, Evidence that T cell activation is required for HIV-1 entry in CD4+ lymphocytes, J. Immunol. 142:773 (1989).
73. N. Mohagheghpour, R. Chakrabarti, B. S. Stein, S. D. Gowda, and E. G. Engleman, Uninfected resting CD4+ T cells require early activation events to fuse with HIV-1 infected T cells, FASEB J. 3:1280 (1989).
74. N. Isakov, M. I. Mally, W. Scholz, and A. Altman, T-lymphocyte activation: the role of protein kinase C and the bifurcating inositol phospholipid signal transduction pathway, Immunol. Rev. 95:89 (1987).
75. S. M. Schnittman, M. C. Psallidopoulos, H. C. Lane, L. Thompson, M. Baseler, F. Massari, C. H. Fox, N. P. Salzman, and A. S. Fauci, The reservoir for HIV-1 in human peripheral blood is a T cell that maintains expression of CD4, Science 245:305 (1989).
76. D. D. Ho, T. Moudgil, and M. Alam, Quantitation of human immunodeficiency virus type 1 in the blood of infected persons, New Engl. J. Med. 321:1621 (1989).
77. R. W. Coombs, A. C. Collier, J.-P. Allain, B. Nikora, M. Leuther, G. F. Gjerset, and L. Corey, Plasma viremia in human immunodeficiency virus infection, New Engl. J. Med. 321:1627 (1989).

## DISCUSSION

Engleman (Stanford): Just to very quickly summarize: all of the data generated by Barry Stein indicate that there is no dependency on acid pH, that fusion is requisite, whether or not it takes place at the surface, or in an endosome. And further, that data which were published in Cell a couple of years ago have been reproduced in normal activated T cells. So, in both the T cell lines and in PHA-activated T cells, these findings hold. I should just add one or two points to this, and that is that everything that was done with PHA-activated T cells has been done with OKT3-activated T cells, so it is not a function of a lectin approximation. The second point that should be made is that it takes very little time after exposure to these stimuli to begin to see evidence of viral entry by this relatively insensitive technique. Others in the lab working with Shantharaj Gowda and Barry have shown that exposure to OKT3 for as little as 15 minutes is sufficient to get virally induced entry as well as cell-to-cell fusion. And perhaps most important, is the observation that Shantharaj pointed out, if you mix fresh resting T cells with a virus envelope-expressing cell line and wait long enough, you do begin to see syncytia formation. But what is most important is that it can be completely blocked by H7 and other agents that will block protein kinase C (PKC) translocation, protein kinase C activation, which is known to be one of the very earliest events in T cell activation. As soon as you pass that point of PKC activation, you cannot block any further, and indeed, these cells do not look activated in the sense that they're not large, they're not incorporating $^3$H-thymidine, etc. There is a caveat, in fact there are at least two caveats to this work: One is that the technique that's used for detection for reverse transcribed viral DNA in the cytoplasm is relatively insensitive, compared to polymerase chain reaction (PCR) amplification. Nonetheless, the difference between entry, or I should say the difference between the appearance of reverse transcribed DNA in activated cells versus resting cells, is manifold, so whether or not one can get some entry of virus that would be detectable by amplification, as opposed to detectable by the kind of technique that Gowda used, is a different issue. Clearly, you can get enhancement by activation. Another caveat is, that while it would appear that activation is clearly necessary for cell-to-cell fusion, we have not proven directly that the appearance of reverse transcribed DNA in activated cells, and not in resting cells, is not due to some other defect. It could conceivably be due to a failure to reverse transcribe viral RNA in resting cells. It's conceivable. However, with respect to the recent reports that claim that you can see HIV in resting cells exposed to virus, if you use PCR-based detection, I would just say that we should be very cautious about interpreting PCR-based detection in a system where you have a detection approach that is so sensitive that if there is perhaps a tenth of a percent of activated T cells in the preparation, you will get obvious entry and I think that one out of a thousand cells in normal individuals is likely to be activated. So, in a sense, using an insensitive technique may be giving us more information than using the terribly sensitive technique of PCR-based amplification. But there are undoubtely people here that feel otherwise, and it's conceivable that either the viral strain or the way the cells were prepared may affect this observation. In any event, I know it's a controversial point, but I just wanted to reiterate our interpretation and why we feel the way we do.

McDougal (Centers for Disease Control): If you take unstimulated cells, incubate them with virus, then strip the surface-bound virus and then stimulate with PHA, do you get a signal? No? We ultimately get infection. The efficiency is less than parallel experiments with activated cells. So it's under conditions of limiting virus input and in a limiting dilution assay with excess virus, and then diluting out the cells to enumerate the frequency of infected cells. Non-activated cells are less efficient. I think what I would argue is whether or not this is a quantitative or qualitative assay.

Engleman: I don't think there's anything that Shantharaj or any of the people in my lab have done that would argue against a quantitative as opposed to a qualitative difference.

McDougal: We have done it with resting T cells, adding virus and then coming back a week or eight days later and still being able to stimulate virus production, and that's following one or several trypsin treatments.

Stein (Stanford): One thing you must consider is, that in any resting cell population that you isolate, you will have a certain percentage that are activated. Then you don't know that the source of the virus that's coming out is not from activation of cells that the virus has already entered. When you further activate them, they produce virus. You don't know the source unless you do in situ hybridization, or some other technique like that.

Engleman: Okay, the other thing that we wanted to emphasize is that prolonged culture in vitro probably gives a sufficient activation signal to get virus inside, in fact, it doesn't take much, but it is interesting that you can block it with PKC. At least you can block the cell-to-cell fusion that you get, and this is easy to reproduce.

Scofield (UCLA): What a beautiful story! What interesting surface markers have appeared on T cells one hour after PHA stimulation?

Engleman: I don't know of any that have been defined using the standard panels of monoclonal antibodies, although we do think that what we're seeing is a membrane event. I don't think any of the monoclonal antibodies that have been defined to date see profound changes in density of expression within an hour. As I said, you can actually get these changes within 15 minutes or so of exposure, so it's a very rapid event.

Scofield: When does class II transcription begin, after activation?

Engleman: Oh, I think it's much later.

Scofield: You think so?

Engleman: I think so. I think the earliest events are calcium flux and PKC activation. And then there's a series of gene activation events that Jerry Crabtree has defined. I think there's a review in Science that looks at that. There may be other changes.

Gowda (Stanford): If you do a few fusion experiments after treating with cycloheximide, you will essentially find the same results. I cannot rule out the possibility of an internal pool of surface protein that may be present.

Engleman: That is, cycloheximide-treated activated cells fuse with the same efficiency as untreated cells.

El-Amad (UC Berkeley): I tend to agree that you would increase the level of expression by stimulation, but not necessarily the infection of the cell. I did experiments on the infection of PBMC which is probably more sensitive than cell lines, where I used resting cells in the patient's serum (5% for minimal toxicity), and both express viral antigens by immunocytochemistry. However, the level of viral expression was one in a thousand in the unstimulated, vs one in ten expressing virus in the stimulated. We also infected the unstimulated cells and after 10 days subjected them to PHA stimulation, and we got high level expression.

Engleman: Yes; I don't know what the details of your first method were. The second one, with regard to using cells in culture for a week to ten days, I think all bets are off, because you can't keep cells in culture for 7 to 10 days and not have this kind of activation taking place. It's a virtual impossibility. And it's well known, and I don't think anyone here would argue, that if you further activate T cells and if you have any integrated virus, you're going to get greatly accentuated virus production in cells that have been activated. I don't think that's in dispute. I'm not sure we're going to resolve this. My particular concern is about using the polymerase chain reaction as the basis for making these conclusions. In the end, I think we come to an almost impossible question. Since there is by definition going to be a small percentage of activated cells in these preparations, and by definition, PCR amplification will detect literally a single copy if it's done a certain way (single copies of virus) I think it's very difficult to use PCR. I know that in the group from Southern California it was said that an initial control was negative, that when they added virus, there was no signal. Even that I would find difficult to understand, since there's usually DNA, some signal from DNA that's contaminating the virus cultures, but it is very difficult.

Pauza (Salk Institute): I think we have to be fair about these experiments though, because Irving Chen's lab is reporting a fundamentally different observation. They're reporting that you infect resting cells and you get strong stop DNA, entirely consistent with the literature on type C viruses from Temin's lab, where a short extension product is made and then the reaction arrests until the cells move out of the $G_0$ phase of the cell cycle. So they have to look by PCR, or some other method, for that very small single-stranded DNA species, and I wonder how you're going to rationalize this between the two experimental systems.

Engleman: Well, I'm not familiar with the data from Chen's lab, so it's a little difficult to comment on data that I haven't heard about. What I'm saying is that anything that uses PCR as a detection method is subject to the problem that if you have any activated cells present in the preparation, you're going to get a positive signal. So it's a tough one to resolve. I don't think it's easily resolved. In any event, I think we would agree that at least there's a very substantial quantitative difference in viral entry between cells that have been activated and cells that have not been deliberately activated. It is fairly important in terms of the concept of so-called latency and therefore, the discussion is not one that's simply picayune.

Galloway (UCSF): Slightly separate question. On some T cells in the presence of phorbol esters, CD4 will internalize and internalize rapidly, in fact down-regulate. On resting T cells with PHA, is the CD4 level on the surface changing rapidly, internalizing, and going down?

Engleman: It changes slightly. Gowda showed a slide of what happens with PHA-exposed T cells, and what you see during the first few hours is a slight decline in the level of CD4 by maybe 5 to 7 percent, and then it goes up over the next 12 to 18 hours to maybe a 10% increase. Eventually, on a fully activated cell, you get a substantial increase in CD4 expression, as you do with a number of other molecules.

Galloway: That's quite different from levels seen in other systems.

Engleman: Yes.

Konopka (UCSF): I have worked with HIV for a short time, but many years with transferrin. We sometimes forget that the receptor-mediated pathway for transferrin is only <u>one</u> pathway, but we have the second pathway which demands the receptor, but is without endocytosis, and this is recognized as the most important pathway for hepatocytes. It seems to me that we should remember that we may have a similar situation with HIV. That we can have endocytosis and fusion and maybe other processes too.

Engleman: Getting back to the entry question, I don't think that it's necessary to dispute the issue of whether or not endocytosis occurs, i.e., that you can see virus in endocytic vesicles. I think the issue that Barry was trying to emphasize, is that you <u>must</u> have fusion, you must have a membrane fusion event to get true infection, and that fusion does not require acidification and therefore, in different cell lines, you will see different amounts of endocytic vesicles with virus in it. But as far as CD4-based entry, entry is dependent upon a membrane fusion event which is pH-independent. I think there's no problem with the possibility of a virus in an endosome fusing and infecting the cell, but it does not require an acid pH environment.

Doms (NIH): I agree with your conclusion there completely. Looking at the rates at which Paramyxoviruses fuse at neutral pH, typically on the order of one-half to two percent of virus particles fuse per minute, which is relatively slow compared to the rate at which many cells internalize plasma membrane, it's even likely that neutral pH fusers like HIV or Paramyxoviruses may find themselves internalized before they have an opportunity to fuse, and since they're not activated by acid pH, they may very well fuse from within endosomes, in addition to the plasma membrane. An interesting question would be to consider the viruses that fuse the plasma membrane: Whether they uncoat as efficiently as viruses which fuse from within endosomes, and whether the virus that fuses from within an endosome finds itself in the center of the cell close to the nucleus. But at least in some cell types, there's a very extensive cytoskeletal network beneath the plasma membrane and there may actually be some physical constraints to uncoating, following fusion directly with the plasma membrane. I have no problem believing that a neutral pH fuser may fuse within an endosome as well as the plasma membrane, given the slow rates at which these things often fuse.

Stein: One experimental observation is incubating the cells at 4° and then washing them about ten log washes (without trypsin treatment), and they wind up with a productive infection. Now you can argue and say there are some virus particles that are left on the cell surface and then endocytosis occurs after rewarming; that's a possibility, but since fusion can occur at 4° and occurs rapidly, there's some presumptive evidence, although not proof, that infection can occur at the plasma membrane. It doesn't have to happen at the level of the endosome. Also, you have to remember that endosomes go through maturation, and that acidification happens relatively rapidly and endosomes become larger and bigger and it takes four or five minutes for them to get to the trans-Golgi network, so you're really not that much closer to the nucleus when viruses really are fusing, because it's probably an initial event. And one other thing I'd like to say, which is a little separate, is the pH independence, which says something very biophysical about the spike glycoprotein. It says something about its conformation. Maybe just the binding to CD4 is enough and you don't have to have a folding of the molecule any further for the hydrophobic domains to be in place, or maybe there is conformation that is pH-independent. I think the other thing we have to be very careful about is comparison to influenza, because there are very big differences between HIV and influenza. The influenza hemagglutinin is cleaved after it leaves the Golgi complex when it gets to the plasma membrane, but that's not the case for the HIV envelope protein. Influenza hemagglutinin is a disulfide-linked molecule, whereas the HIV envelope protein isn't. It's also acid-dependent, where the HIV envelope protein isn't. So I think when we start comparing one virus system to the other, we have to have a lot of latitude in terms of what we accept as dogma and what is really controversy.

Davis (Medical Research Institute, San Francisco): I'm a little bit concerned about the extension of results from failure to detect the reverse-transcribed DNA to a conclusion about whether fusion is the relevant event. Have you performed any experiments to look directly at, let's say, input RNA in the cytoplasm cells, or some other earlier process?

Engleman: No, I think that's a fair criticism. As I indicated before, we have no direct evidence that the fusion phenomenon is responsible for the failure to detect reverse-transcribed DNA in our system. It could be either a failure to unravel or a failure to reverse transcribe. We favor the fusion because of the cell-to-cell fusion which does not take place and is enhanced tremendously by T cell activation.

Fuller (University of Michigan): Just to make a point about understanding the entry process of HIV, the fact that it does seem to be pH-independent says that there must be some other way that the virus is able to protect its fusogen until it's at the point that it's required, and I think that's the central question that we can look at for Herpes viruses as well as for HIV. What is the mechanism by which this virus is able to allow its fusogen to remain protected until it's time for it to do what it does? And I think that's the central question that maybe we'll talk a little bit more about tomorrow. But, it is obviously different from those viruses that go in by a pH-dependent mechanism which have the acidification process to change its conformation. I hope someone addresses that here.

Engleman: Well, I think Sodroski will have all the answers tomorrow. I should mention to you, for those of you that have not, do not, play with T cell activation and this kind of phenomenon, don't use Concanavalin A. Con A is just as commonly used as PHA to activate T cells, but Con A won't work. Con A-activated T cells, at least if you did it the way we did it several years ago, is not readily infectable because Con A binds to the viral carbohydrate, and that was a bit of a surprise then. Today it's common knowledge.

Sodroski (Harvard): I guess I have a question and it's sort of a comment also: it's been reported that gp120 itself can activate T cells, and I'm just wondering how much gp120 do you really have to put on cells to see an activation and is that what might be occurring over this 24-hour period, for example?

Engleman: It's a possibility. It's just that the reports of gp120-induced activation involve really enormous concentrations, as I recall, microgram quantities; but it's a possibility. Another recently reported phenomenon relating to cell-to-cell fusion is that anti-LFA1 monoclonal antibodies apparently can block it. This is very logical, but I would point out to you that many labs looked for that a long time ago, ours included, and we did not observe that anti-LFA1a or b antibodies inhibited syncytia formation, and I assure you we were very interested in that point. There are other antibodies that do, which we're not quite ready to talk about yet, but in our hands, anti-LFA1 did not.

Woodle (Liposome Technology, Inc.): It's my understanding that CD4 expression on a number of other cells besides human cells can give rise to binding and perhaps fusion, but not infection. Do you have any experimental grasp of that sort of situation?

Engleman: I think it's fair to say that you get binding with CD4 expressed in any cell type. The data is primarily from the mouse cell data, and I'm sure that either Steve McDougal or Jay Levy or someone else can address that better than I. But the implication is that you need at least one other protein, either on the surface or inside the cell.

Mills (UCSF): Do you get reverse transcription in the mouse cells? Can you find cDNA, in an analogous sort of situation, in the CD4-positive mouse cells?

Levy (UCSF): We have not. If you transfect, we get low levels of viral replication.

Mills: But in mouse cells that express human CD4, if you expose them to virus, do you get viral cDNA?

Levy: I can't answer that, the only thing we know is we do not get infectious virus.

Engleman: The fact is, though, that there is a rabbit model that has been described at some length, and normal rabbits are apparently susceptible in vivo to some form of HIV-1 infection. This model has been described in the literature and it was discussed at length in Montreal. The rabbit CD4 molecule can't be that close to the human CD4 molecule, so I think the evidence is building that there are other ways of getting the virus inside cells besides the work of Jay Levy and others, so I would be surprised that if you bathed the mouse CD4-positive cell long enough, you wouldn't get some form of low level infection.

Düzgünes: I don't know what the energy levels of both the audience and the speakers are, but the motto of the workshop is "We Shall Leave No Stone Unturned," so if there are any questions at all, I would like to bring them up at this point for a more general discussion involving all the presentations that we have heard today.

Engleman: The only problem is that there are other speakers who are going to address basically some of these same questions, so it's really not fair, but it's okay with me.

Pauza (Salk Institute): Does PHA bind to CD4?

Engleman: I don't think so. No. Binds to CD3 though, but not CD4. Remember I said the same phenomenology can be demonstrated with other stimuli, virtually any stimulus except for Con-A, and other high-mannose binding lectins.

El-Amad: My question is for Dr. Ho. Have you looked at the hepatoma cell lines for HBV DNA, using PCR, as a co-factor for HIV expression? I wasn't aware that hepatomas are infected by HIV alone.

Ho (UCLA): In three of the five hepatoma cell lines, we don't have to use PCR techniques to see HBV. They're actively expressing hepatitis B surface antigen, but in two of them they're not expressing the surface antigen.

However, they do carry the HBV genome. So all five of them have HBV in them. Interestingly, with the three that express surface antigen, there is an inverse relationship between p24 antigen expression and HBV surface antigen expression. As HIV replication goes up, the HBV surface antigen production goes down. We have not done any studies to address possible interactions between those two viruses.

Fuller: This is a general question to anyone here. In terms of tropism of the immunodeficiency viruses, I think there was a little bit mentioned earlier about HIV-1 and HIV-2 and SIV by Dr. Doms, but is there anything known about entry and tropism, and why certain viruses have different tropisms for different cells regarding entry?

Stein: I am sort of a newcomer to tropism, even though I walked into it by accident. When you look at the literature everybody addresses virus replication, they look at entry, but they don't look at the entry events which would approximate unraveling or linear duplex formation, and I would like to bring attention to the FT1 locus in mice, where you have restrictive replication. For MuLV, depending on what the specific fibroblast type is, there's a specific gene that dictates the restriction and it maps to the gag region of MuLV. It looks like we have two variants, populations, which may or may not be an FT1 locus, but may be similar to some other locus that may be restricting the formation of a linear duplex form in integrated DNA. We need to look at that more closely, and we have to do complementation studies to rule out the possibility that the difference in the amount of linear duplex that we're seeing between variants and parents, are not due to attenuated fusion, by doing complementation with envelope regions. But my belief, based on the data that we have, is that it's probably at the level of nucleic acid processing and we need to think about this when we think about tropism, along with envelope-mediated events such as binding and fusion.

Levy: Well, the question that you posed about tropism is the whole reason for this meeting. We really don't know why a virus will selectively go into a T cell or a macrophage. It doesn't appear to be binding the CD4. That was the leading theme when we started presenting the fact that you could infect brain cells without CD4. No-one wanted to believe that, and I can tell you that, in talking with companies, they still don't want to believe it, and claim CD4 is the only way you're getting a virus in. I think all of us here have worked enough to know that that's not the only way. But it certainly says to us that if it's there, it also doesn't distinguish tropisms. So we have to move to the next step, which is after attachment, fusion, or if you don't have attachment by CD4, what other mechanisms? I don't think the macrophage has a separate attachment receptor. I think it has a separate event of fusion, perhaps with gp41. And that will be true of fibroblasts, and true of glial cells and hepatoma cells and neuroblastoma cells. So the key element in the research should be, what is the event that occurs after attachment? And the question then is: are the gp41s that different? Well most of you know that gp41s are highly conserved. What would happen if you put a gp41 on a virus that is non-macrophage-tropic from a virus that is, will you get a change? I'm afraid you won't, because I think the conformational change is important and the molecule probably has to be fairly intact. But that's where I would put the direction in terms of the early event tropism. Coming from what Barry said, I came from FD1 locus and when we were doing the cut-and-paste experiments, I took the side of the 5' end just to be ornery, although I knew it would be the 3' end, because of the gag proteins. That says that the virus gets in, but it just can't be detected as an infectious particle because you don't see it replicated. So the question then says, does it integrate? And you may not see that by Southern blot. You may, if it's not a highly prevalent event, and yet it may be important. And it may be seen only with PCR and yet that level, that technique, is also possibly not that sensitive. What is the definition of a tropism? Most nonvirologists in the field look on tropism as "can you get the virus out". And they report it that way. We cannot infect fibroblasts. When we presented our fibroblast data, first people said "residual virus". You have to trypsinize and get rid of it, and then, if the virus doesn't come out it doesn't mean the virus isn't there. I mean, latency is an example of that. So we have to recognize that many different ways for infecting may be present. The virus gets in, and we have to have good techniques for finding the virus there, and PCR may be it, but the best is in situ PCR if it's ever developed. And then the question is, if it is in the cell, and this gets to be a question of pathogenesis, does the integrated virus produce anything that could cause a disruption in cell function? And I found that issue very interesting and very important. We've already discussed some of this. I mean, what's the calcium influx of cells that you think are infected with HIV? They're making virus, but have they changed the permeability so that you can really see differences in electrophysiologic state? Maybe, as I commented, glial cells aren't working right, so dementia develops because you got virus there but you don't see virus replicating, you don't see a lot of viral RNA. So I find it fascinating to watch the field now, and you have a lot of people in the CD4 field. Terrific. Let's see what happens there. I think you've got enough exceptions to say CD4 plus something is occurring. And then there are individuals who now want to concentrate on the non-CD4 mechanism. Which may actually be the mechanism. I mean, David Ho has his hepatomas and neuroblastoma cells; the way you enter them may be inefficient because you don't have CD4 there. I bet if you put CD4 on you have a terrifically good infection. That then becomes a very powerful approach to look at tropism. Difficult to work in because the efficiency is so poor. And then the third one is this antibody enhancement, which we've discussed, which appears to be FcIIIR in the macrophage and maybe another Fc receptor in another cell. Maybe another mechanism in fibroblasts and what we would argue, coming from this field, would be, don't put all your eggs in the same basket, don't go after one thing. In fact, I don't know whether any receptor for virus entry is that important. More important, possibly, is the fusogenic effects of the cell or the cell-to-cell contact, because that's where I think the virus spreads. So if we can tie the two together, so much the better. After this lengthy talk, my main concern is that tropism is what we need to understand in order to get a good antiviral drug. And that's why CD4 is being used, and we need to find a more common denominator, or define all the denominators, so that we can go after this and most important is to stop the spread of cell-to-cell contact, so when you're looking at tropism and you think you've got something, make sure you can block it if you bring two cells together.

Düzgünes: Thank you very much, Jay. One comment from Dr. Sodroski:

Sodroski: Just to expand on something Jay said. Fusion in fact may be very important and it may be the important determinant, I think, for some of the tropisms that have been seen. I also think we shouldn't necessarily think only of gp41 as the fusogen, because I think, as you mentioned, it's very highly conserved and it's probably premature to think that there are going to be changes in the gp41 amino terminus or any other ectodomain of gp41 that is going to confer tropism. I think we also have to remember that there is a lot of evidence that mutations, even in hypervariable regions of gp120 and antibodies directed against the variable region 3 of gp120, can also affect post-CD4 binding events and virus entry, namely events that really come under the umbrella of fusion at this point, since we don't really know how to distinguish what happens after receptor binding. There may be multiple steps, but everything right now is just classified as fusion. In fact, I think that in addition to thinking about gp41 as a fusogen, we also have to remember that probably gp120 plays an important role in initiating fusion processes and that virus variants, for example the different tropic variants, may in fact map to regions in gp120 that are important for initiating fusion events as well. The other point is that, as far as we've been able to see, fusion itself has always required the presence of the CD4 molecule, presumably to activate some sort of conformational change in the envelope glycoproteins, so if gp41 in and of itself can allow virus entry into non-CD4-bearing cells, one would have to wonder what takes the place of that potential conformational activation step of gp120. So, these are just questions that no-one has answers for at this point, but they are things to bear in mind when we're talking about deciphering tropism and looking for "second receptors." Those second receptors may not in fact be the gp41 at all but may involve other cellular molecules that are actually taken up and incorporated into virions.

# HIV-1 INFECTION OF A NON-CD4-EXPRESSING VARIANT OF HUT-78 CELLS: LACK OF INHIBITION BY LEU3A ANTIBODIES AND ENHANCEMENT BY CATIONIC DOTMA LIPOSOMES *

Krystyna Konopka‡, Brian R. Davis† and Nejat Düzgüneş‡ # §

‡Cancer Research Institute and #Department of Pharmaceutical Chemistry, University of California, San Francisco, California 94143; §Department of Microbiology, University of the Pacific, School of Dentistry, San Francisco, CA 94115; †Medical Research Institute of San Francisco at Pacific Presbyterian Medical Center, San Francisco, CA 94115

## INTRODUCTION

CD4 and CD8 are membrane glycoproteins that have been implicated in recognition of MHC class-II (HLA-DR, DQ, DP and I-A/I-E) and class-I (HLA-A,B, C and H-2) antigens, respectively. CD4 and CD8 are expressed on functionally distinct populations of mature peripheral T lymphocytes; CD4 on helper/inducer cells and CD8 on cytotoxic/supressor cells (Reinherz and Schlossman, 1980). The CD4 molecule is expressed as a monomer on the cell surface of T lymphocytes and some cells of the macrophage/monocytes lineage, and is a transmembrane glycoprotein of ~ 55 kD with homology to members of the immunoglobulin (Ig) supergene family. It has four external Ig variable-like domains ($V_1$ - $V_4$), one transmembrane domain and a cytoplasmic tail of 40 residues (Maddon et al., 1985). The precise role of the CD4 molecule in T cell function is unknown. It has been proposed that CD4 may increase the avidity of the T cell receptor (TcR) for its ligand(s) by directly binding to monomorphic determinants of class II MHC molecules on target cells. Transfection of a cDNA encoding the human CD4 molecule into antigen-responsive T-cell hybridomas greatly enhanced the cells' responsiveness to antigenic stimulation (Doyle and Strominger, 1987; Gay et al., 1987; Sleckman et al., 1987). In addition, CD4 may associate with the TcR as part of the antigen recognition process (Janeway et al.,1987; Saizawa et al., 1987) and may be directly involved in the resultant signal transduction (Emmerich et al., 1987; Carrel et al., 1988), possibly in concert with a lymphocyte-specific protein tyrosine kinase (reviewed by Robey and Axel, 1990).

* Address for correspondence: Dr. N. Düzgüneş, Department of Microbiology, University of the Pacific, School of Dentistry, San Francisco, CA 94115

The cell surface expression of CD4 has been shown to be down-regulated following treatment with phorbol esters (Acres et al., 1986; Hoxie et al., 1986; Clapham et al., 1987) or during antigenic stimulation (Rivas et al.,1988; Habeshaw and Dalgleish, 1989).

CD4 also functions as a receptor for the human immunodeficiency virus type 1 (HIV-1) envelope glycoprotein gp120, both on T-lymphocytes (Dalgleish et al. 1984; Klatzman et al., 1984) and on monocytes and macrophages (Gartner et al., 1986; Ho et al., 1986; Clapham et al., 1987). The sensitivity of the cells to HIV-1 infection is associated with the expression of the CD4 protein on their surface (Sodroski et al., 1986). Recent studies have shown, however, that certain glioma, neuronal, fibroblastic and myoblastic cells in culture can be infected by HIV-1, although at a relatively low plating efficiency compared with T-lymphoblasts (Cheng-Mayer et al., 1987; Dewhurst et al., 1987; Clapham et al., 1989; Harouse et al., 1989; Tateno et al., 1989; Li et al., 1990). Most of these cells appear not to express cell-surface CD4 as measured by fluorescence, but some of the susceptible cell lines do have low levels of CD4 mRNA (Weber et al., 1989). Anti-CD4 antibodies or neutralizing doses of soluble recombinant CD4 (Clapham et al., 1989; Weber et al., 1989) do not reduce the low level of infectivity by HIV-1 seen for these cell types (Weiss et al., 1989), suggesting that the CD4 antigen is not required for infection of certain non-haematopoietic cells by HIV-1. Furthermore, expression of the CD4 molecule on the cell surface may not be sufficient for HIV-1 infection (Maddon et al., 1986; Chesebro et al., 1990).

Here we report our observations on the interaction of HIV-1 with a non-CD4-expressing variant of the CD4-positive parental HUT-78 cells, which we have termed "HUT-90." We have found that these cells could be infected by HIV-1, as determined by the production of p24 (gag) in the cell culture supernatant and in situ, despite the absence of detectable CD4 antigen on the cell surface. The infection was not inhibited by Leu3a antibodies. We have reported recently that liposomes composed of the positively charged synthetic lipid N-2,3-(dioleyloxy) propyl]-N,N,N-trimethylammonium chloride (DOTMA) enhance the infectivity of HIV-1 and that the enhancement correlates with the expression of CD4 on the cell surface (Konopka et al., 1990a,b; Düzgüneş et al., 1991). Surprisingly, infection of HUT-90 cells by HIV-1 was enhanced by DOTMA liposomes. We have also found that freezing and thawing in appropriate media altered the expression of CD4 on the surface of the HUT-90 cells.

## METHODS

Chemicals. N[2,3(Dioleyloxy)propyl]-N,N,N-trimethylammonium chloride (DOTMA) was synthesized as described by Felgner et al. (1987), and stored in ethanol under argon at -70°C. Anti Leu2, Leu3, Leu3a and Leu4 antibodies were obtained from Becton Dickinson (San Jose, CA).

Cell lines. The human cutaneous T cell lymphoma cell line (HUT-78) was obtained from the American Type Culture Collection. "HUT-90" cells were derived from the parental HUT-78 cell line after two years of passaging. Both HUT-78 and HUT-90 cell lines shared the identical MHC class I antigens (A1,

Aw68, Bw62, Cw3). While HUT-78 cells expressed the class II MHC antigens DR4, DRw53, and DQw3, the HUT-90 cells expressed DR4 at only lower level, and failed to express DRw53 and DQw3. The chronically infected H9/HTLV-$III_B$ cells were used as a positive control for the determination of viral p24 antigen inside the cells. The CD4+ human lymphoblastoid cell line, A3.01, was obtained from Thomas Folks (Centers for Disease Control, Atlanta, GA). Cells were cultured in RPMI-1640 medium supplemented with 10% (v/v) heat-inactivated fetal bovine serum (FBS), penicillin (50 units/ml), streptomycin (50 µg/ml), and L-glutamine (2 mM), at 37°C (7% $CO_2$).

Virus. The virus and infected cell lines were handled in P-2 and P-3 facilities as recommended by the Centers for Disease Control (1988). A standard laboratory strain of HIV-1 (LAV) was originally provided by Thomas Folks and was propagated in A3.01 cells. The supernatant containing the virus was harvested at times of peak reverse transcriptase activity, aliquoted and stored at -80°C. The reverse transcriptase activity of the virus stock was 5.1 x $10^5$ cpm/ml supernatant, as detected by the method of Hoffman et al. (1985). The ELISA-reactive p24 protein in the supernatant yielded approximately 1.0 µg/ml. Viral inocula were standardized by their content of p24 antigen, and cultures were infected with 1 or 2 ng of viral p24 per $10^6$ cells in all of the experiments.

Preparation of liposomes. Large unilamellar vesicles composed of DOTMA were prepared in 150 mM NaCl, 10 mM N-[*Tris* (hydroxymethyl) methyl]-2-aminoethane sulfonic acid (TES) (Sigma, St. Louis, MO, and Calbiochem, La Jolla, CA), pH 7.5, by hydrating dried lipid (Bangham et al., 1965). Vesicles were extruded under argon three times through double polycarbonate membranes of 0.2 µm pore diameter (Poretics, Inc., Livermore, CA) in a Lipex Biomembranes (Vancouver, BC) high pressure extrusion cell, to achieve a uniform size distribution of vesicles (Szoka et al., 1980). This procedure results in the production of predominantly unilamellar vesicles (Düzgüneş et al.,1983; Hope et al., 1985).

Detection of CD4, CD8, and CD3. The presence of cell surface CD4, CD8, and CD3 proteins was examined by direct immunofluorescence assay, using FITC-conjugated Leu3, Leu2, and Leu4, respectively, with FITC-conjugated mouse IgG as a control. The fluorescence distribution was analyzed with a fluorescence-activated cell sorter (FACScan; Becton Dickinson, Mountain View, CA).

HIV infection of cells. 2 x $10^7$ cells/ml were incubated with HIV-1 for 2 h at 37°C to permit adsorption of virus particles and then washed three times to remove unbound virus. Cells were then resuspended at 5 x $10^5$ cells/ml in RPMI-1640/10% FBS and cultured at 37°C in 25 $cm^2$ culture flasks and maintained in 8 or 10 ml of medium. Control cells were similarly treated but not exposed to the virus. Samples of the supernatant (1 ml ) were collected every 3 or 4 days for subsequent screening by enzyme-linked immunosorbent assay (ELISA) for HIV p24 antigen.

HIV-1 (0.4 µg viral p24/ml) was pre-incubated with DOTMA liposomes for 2 h at 37°C, and then diluted with a cell suspension to give 1 or 2 ng of viral p24 protein per $10^6$ cells. After virus adsorption for 2 h at 37°C, the cells were washed three times, resuspended at 5 x $10^5$ cells/ml and cultured in 25 $cm^2$ culture flasks.

Following incubation, the number of viable cells was determined by trypan blue exclusion, and p24 antigen was measured in duplicate in the supernatant.

Detection of HIV infection. HIV-specific antigens were detected in cell culture supernatants using an HIV-1 p24 antigen capture assay. The expression of HIV antigen in the infected and mock-infected cells was determined using monoclonal mouse anti-p24 and soluble complexes of calf intestinal alkaline phosphatase and mouse monoclonal anti-alkaline phosphatase antibodies (APAAP). After the cells were washed with phosphate-buffered saline (PBS, Zymed Laboratories Inc., South San Francisco, CA), they were suspended in PBS, placed on 10-well slides and fixed for 3-5 min in acetone. The slides were overlayed with a mouse monoclonal antibody to HIV-1 p24 protein, rabbit anti-mouse immunoglobulins, and APAAP (Dako Corporation, Santa Barbara, CA). Staining was performed using the Alkaline Substrate Kit (Vector Laboratories, Burlingame, CA). The uninfected cells and H9/HTLV-$III_B$ cells were used as negative and positive controls, respectively.

p24 antigen capture assay. Cell-free supernatant fluids were assayed in duplicate by an HIV-1 p24 antigen capture assay, with two mouse monoclonal antibodies specific for HIV p24 gag protein recognizing different epitopes. ELISA plates (96 well; Corning, Palo Alto, CA) were coated overnight with the capture antibody, incubated with 0.1% bovine serum albumin in PBS for 1 h at room temperature and washed three times with PBS-Tween 20 (0.05%). Next, 100 µl of supernatant was added, and samples were incubated for 2 h at 37°C. The biotinylated detector antibody was subsequently added and after 1 h incubation at 37°C the antibody was bound with avidin-horseradish peroxidase. The absorbance of the color that develops from the reaction of the peroxidase with hydrogen peroxide in the presence of orthophenylene-diamine (OPD; Dakopatts, Glostrup, Denmark) was measured at 490 nm, using a Bio-Tek (Burlington, VT) microplate reader. The p24 antigen was quantitated from serial dilution of a positive control preparation containing a known amount of antigen. A culture supernatant containing a known amount of p24, calibrated with the lyophilized Antigen Reagent (Coulter™ HIV Ag Assay; Coulter Immunology, Hialeah, Fl), was used as a standard. The sensitivity of detection was 38 pg of p24 per ml of cell culture supernatant. Statistical comparisons were performed using a Student two-sided unpaired t-test.

Blocking experiments. HUT-78 and HUT-90 cells ($2x10^7$ cells/ml) were pre-incubated with the anti-CD4 antibody Leu3a at a concentration of 10 µg/ml for 1 h at 37°C (Weber et al., 1989). This antibody has been shown previously to block HIV infection of T cells (Sattentau et al., 1986). Both the cells pre-incubated with anti-Leu3a and untreated cells were then exposed to HIV-1 at 2 ng viral p24 per $10^6$ cells for 2 h at 37°C. The cells were then washed three times with RPMI medium, and cultured for 14 days.

## RESULTS

HUT-90 cells, while lacking surface CD4 expression, are infectable with HIV-1. FACS analysis indicated that 79% of the parental HUT-78 cells were CD4+ (Fig. 1 A). The surface CD4 antigen could not be detected in the variant HUT-90, derived from the parental HUT-78 cell line after two years of passaging in RPMI/10% FBS medium (Fig. 1 B).

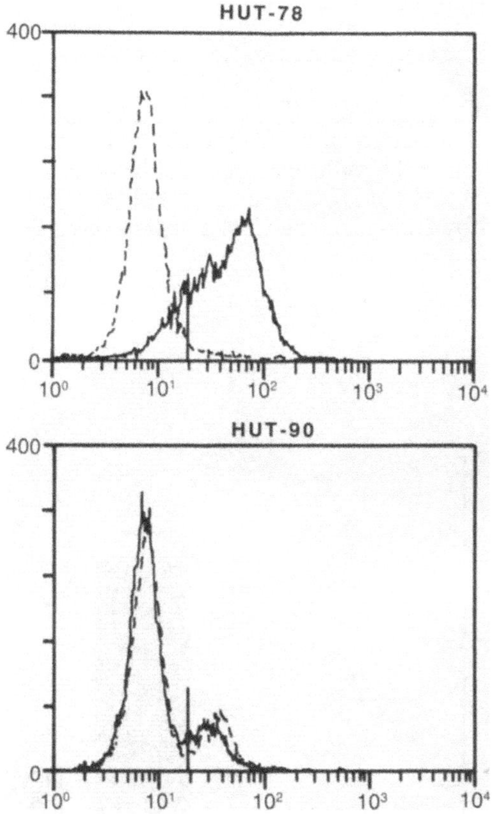


Figure 1. Immunofluorescence cytometry of HUT-78 and HUT-90 cells. The expression of cell surface antigens was determined by direct immunofluorescence assay as described in Methods. Results for a nonreactive antibody (mouse IgG) (dotted line) are superimposed on results obtained with monoclonal antibody against CD4 (anti-Leu3) (solid line).

The CD8 antigen was not expressed on the surface of HUT-78 and HUT-90 cells (data not shown). FACS analysis indicated that 85% and 8% of the HUT-78 and HUT-90 cells, respectively, expressed CD3, the antigen associated with the αβ heterodimer in the functional TcR (Allison and Lanier, 1987) (data not shown). MHC typing of the HUT-78 and HUT-90 cell lines demonstrated identical class I MHC phenotypes; in contrast, the class II MHC antigens for HUT-90 cells showed a decrease in DR and DQ antigens.

Virus production in the HUT-90 cell line was sufficient to be directly detected in the cell culture supernatant, despite the absence of detectable CD4 antigen on the cell surface. Infection was also followed by immunohistochemical staining (APAAP assay) of infected and mock-infected cells. The expression of p24 antigen in HIV-1 infected cells correlated well with the production of p24 antigen determined in the supernatant (data not shown).

Anti Leu3a antibody inhibits HIV-1 infection of HUT-78 but not HUT-90 cells. Although we did not detect any CD4 protein by flow cytometry, it is possible that low-level expression of the molecule could not be detected by this technique. If CD4 is present on the surface of HUT-90 cells at below detectable levels, then the infection of these cells by HIV-1 via the CD4 pathway should still be sensitive to

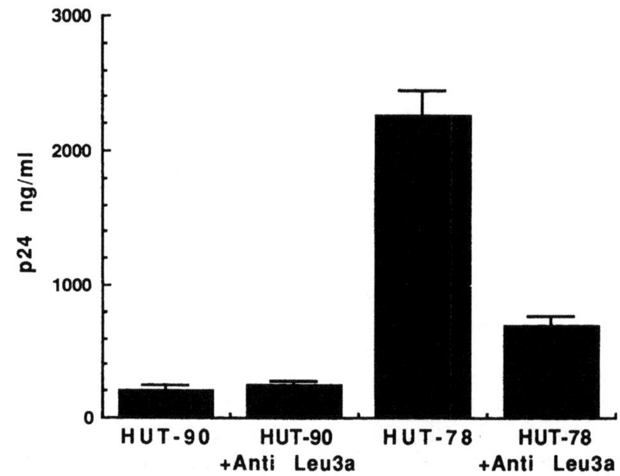


Figure 2. Effect of the anti-CD4 antibody, anti-Leu3a, on HIV-1 infection of HUT-78 and HUT-90 cells. Cells were pre-incubated with anti-Leu3a and then exposed to HIV-1 as described in Methods. The p24 antigen was detected in cell culture supernatants at 14 days post-infection. Data represent mean values ± SD (n=4). Virus negative controls yielded p24 values ≤ 0.038 ng/ml.

blockade by the anti-CD4 monoclonal antibody anti-Leu3a (McDougal et al., 1986).

The results presented in Figure 2 indicate that pre-incubation of HUT-78 cells with anti-Leu3a inhibited the production of p24 protein by approximately 70% after 14 days of culture. In contrast, anti-Leu3a antibody did not block the infection HUT-90 cells by HIV-1, as determined by the p24 antigen capture assay. It appears, therefore, that HIV-1 infection of HUT-90 cells occurs independent of CD4 as the cellular receptor.

DOTMA liposomes enhance HIV-1 infection of HUT-90 cells. The time-course of viral replication in HUT-90 cells is shown in Figure 3 and Table 1. Exposure of HUT-90 cells to untreated virus produced low levels of extra-cellular virus by 7 days post-infection. A pronounced increase in the production of p24 was observed two weeks after inoculation. When HUT-90 cells were inoculated with HIV-1 pre-incubated with 10 μM DOTMA, the production of viral proteins was increased significantly (Fig. 3 and Table 1). Significant levels of p24 could already be detected in the supernatant of 7 day cultures, and at 11 days the p24 value was 10-fold higher than the untreated controls.

Freezing and thawing of HUT-90 cells alters the expression of cell-surface CD4 and CD3 proteins. HUT-78 cells (79% $CD4^+$) and the $CD4^-$ HUT-90 variant were frozen in RPMI/10% FBS medium (90%) plus dimethylsulfoxide (10%), and stored for two months in liquid nitrogen vapor phase (-100°C). The cells were

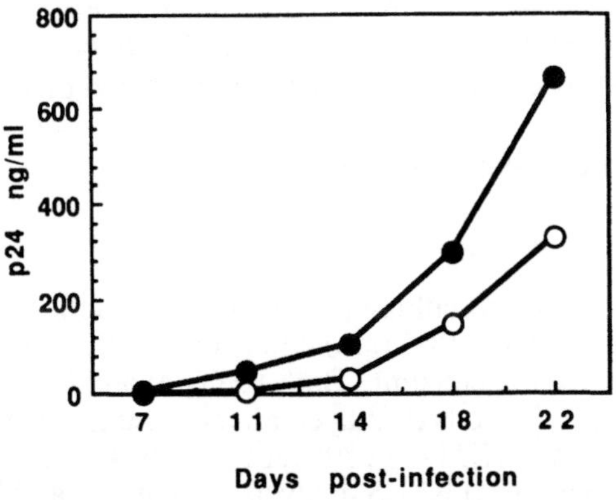


Figure 3. Kinetics of p24 antigen production in culture supernatant of HUT-90 cells infected with HIV-1 pre-incubated with DOTMA liposomes (10 μM lipid) (•) and with untreated virus (o). After a 2 h incubation with the virus, the cells were incubated up to 22 days. Supernatants were collected and assayed for p24 antigen. Data represent mean values (n=4).

Table 1. Effect of pre-incubation of HIV-1 with DOTMA liposomes on viral p24 levels in the supernatant of HUT-90 cells.

| | TREATMENT | |
|---|---|---|
| | DOTMA-TREATED HIV[a] | UNTREATED HIV[b] |
| DAYS POST-INFECTION | p24 (ng/ml)[c] | |
| 7 | 4.8 ± 0.32 | 0.6 ± 0.06 |
| 11 | 47.1 ± 6.4 | 4.7 ± 0.76 |
| 14 | 104.6 ± 12.2 | 28.6 ± 2.63 |
| 18 | 297.0 ± 11.8 | 144.3 ± 10.1 |
| 22 | 663.5 ± 171.5 | 325.5 ± 19.1 |

[a] HUT-90 cells were infected with HIV-1 (2 ng viral p24 per $10^6$ cells) pre-incubated with DOTMA liposomes (10 µM lipid) for 2 hr at 37°C. After absorption of virus (2 hr at 37°C) cells were washed and resuspended at 5 x $10^5$ cells per ml in 25-$cm^2$ culture flasks and maintained in 10 ml of RPMI/ 10% FBS medium. One ml samples of supernatant were collected in duplicate at the indicated time points and assayed for p24 antigen.

[b] HUT-90 cells were infected with untreated virus and then maintained as above.

[c] Data represent mean values ± SD (n=4).

then thawed according to the cell culture propagation instructions (NIAID, AIDS Research and Reference Reagent Program) and split 1:6 every three or four days to give a concentration of 2 x $10^6$ cells/ml. Both HUT-90 and HUT-78 cells formed clusters and clumps. HUT-90 cells displayed an increasing tendency to adhere to the bottom of the flask, while HUT-78 cells remained in suspension. After four weeks of culture (eight passages) FACS analysis indicated that 24% and 15% of HUT-90 cells were $CD4^+$ and $CD3^+$, respectively. A marked decrease in expression of the two antigens occurred when these cells were tested again after passaging twice a week for the following two months. The CD4 antigen could not be detected while 8% of the cells were $CD3^+$. These HUT-90 cells tested continuously $CD4^-$ during next four months of culture. Freezing did not influence the level of expression of CD3 antigen on the surface of HUT-78 cells, although it caused a transient decrease in CD4 expression: 19% of the cells were CD4+ after four weeks of culture (eight passages), while 60% of the cells displayed the CD4 antigen after passaging for the following two months.

## DISCUSSION

Our previous work had shown that the infectivity of HIV-1 was significantly enhanced by DOTMA liposomes and that the sensitivity of A3.01, H9 and K562 cells to the DOTMA-mediated enhancement correlated with the expression of CD4 on their surface (Konopka et al., 1990a,b; Düzgünes et al., 1991). The enhancement was most pronounced in the highly CD4+ A3.01 cells, and lower in H9 cells, while the CD4- K562 cells were not detectably infected. In A3.01 cells, Leu3a antibody blocked 93% of the infectivity of HIV-1 in the presence of DOTMA, indicating that the enhancement of infectivity was largely via the CD4 receptor (Konopka et al., 1990a,b).

In this study, we examined the infectivity of HIV-1 pre-incubated with DOTMA liposomes in HUT-90 cells. After a two year period of passaging HUT-78 cells, a variant appeared in which the expression of CD4 on the cell surface had been down-regulated, and the surface CD4 antigen could not be detected by FACS analysis. These cells, designated HUT-90, were CD8- and slightly CD3+. The expression of CD4 antigen could be transiently increased by 24% after freezing and thawing, indicating that the HUT-90 cells contain the gene for CD4. In contrast, freezing and thawing caused a transient decrease by 60% in the expression of CD4 on the surface of HUT-78 cells. It is of interest to note that simultaneous changes were observed in the expression of CD4 and CD3 on the surface of HUT-90 cells but not on the HUT-78 cell surface. The results suggest some interrelation between the expression of CD4 and CD3 molecules in HUT-90 cells. Exposure of T cells to phorbol ester similarly modulates the cell surface expression of the CD4 and CD3 molecules (Habeshaw and Dalgleish, 1989). This modulation appears to involve the phosphorylation of serine and tyrosine residues on CD3 (reviewed by Allison and Lanier, 1987) and a serine residue on the cytoplasmic domain of CD4 (Acres et al., 1986). Antibody directed against nonpolymorphic determinants within the TcR/CD3 complex significantly reduce the surface expression of CD4, indicating a physical association between the CD4 molecule and the TcR complex in activated T cells (Saizawa et al., 1987; Rivas et al.,1988).

Although the HUT-90 cell line does not express detectable CD4, it can be infected by HIV-1. It is possible that these cells have a very low concentration of the CD4 antigen as the receptor for HIV-1 entry. The low virus productivity of the cells may be related to the very low level of CD4 expression. Nevertheless, HIV-1 infection of HUT-90 cells could not be inhibited by Leu3a antibodies, suggesting the absence of CD4 on the cell surface. Blocking of infection with Leu3a antibody has been taken as an indication that entry occurs predominantly via the CD4 receptor (Sattentau et. al., 1986; Weber et al., 1989; Werner et al., 1990). This antibody inhibits infection of CD4+ lymphoid cells by binding to or near an epitope of the CD4 molecule involved in binding to gp120, and blocking the attachment of the virus to its cellular receptor (McDougal et al., 1986). Some HIV-susceptible brain cells contain mRNA for CD4, but infection is not blocked by Leu3a antibodies (Weber et. al.,1989). Thus, HIV infection of these cells does not appear to involve CD4 as the cellular receptor. Weber et al. (1989) have suggested that a non-specific mechanism may be involved, since higher doses of virus were required to infect these cells, and that a second, low-affinity, receptor for HIV entry may be used. The observations that both murine cells (Maddon et al., 1986) and human brain and skin cells (Chesebro et al., 1990) transfected with human CD4 bind HIV, but are not infected, due to lack of fusion between the HIV envelope and the cell membrane, suggest that a second crucial step, possibly a second receptor, may be needed for infection. Tang and Levy (1990) have shown that the ability of $CD4^+$ cells to undergo fusion with HIV-infected cells is diminished upon treatment with trypsin, although the CD4 protein remains on the cell surface, and have suggested that trypsin may eliminate a specific "fusion receptor." Our studies with liposomes as target membranes for simian immunodeficiency virus or HIV-1, have indicated that the viruses can fuse with certain phospholipid membranes not containing the CD4 receptor, and that the rate of fusion increases upon lowering the pH or increasing the calcium ion concentration (Larsen et al., 1990; Düzgünes et al., 1991). Thus, it is possible that the expression of certain phospholipids on the outer monolayer of the plasma membrane bilayer may also mediate the fusion of HIV-1 with the membrane.

The question of whether the enhanced infection of HUT-90 cells observed with DOTMA-pre-treated virus occurs via the CD4 receptor expressed at sub-threshold levels, or via an alternative mechanism (for example, by direct fusion of the virus with the cell membrane lipid bilayer or a second receptor for HIV entry), is being investigated.

We have observed that long term passaging as well as freezing and thawing can alter the expression of CD4 on the surface of HUT-78 cells; this emphasizes the necessity to monitor frequently the cell surface antigens of cultured cells during propagation or after storage.

## ACKNOWLEDGMENTS

We thank Mr. Norman Dyson and Syntex Research for the gift of DOTMA, Mr. Harold Legg for the reverse transcriptase assay, Ms. Jane Wang and Ms. Diana Flasher for FACS analysis, and Dr. Demetrios Papahadjopoulos for the use of his laboratory facilities. This work was supported by a grant from Liposome Technology, Inc. (N.D., K.K.), grant AI-25534 (N.D., K.K.) from the National Institute of Allergy and Infectious Diseases, grant HL-42105 (B.R.D.) from the National Heart, Lung, and Blood Institute, and grant 86SF017 from the Universitywide AIDS Research Program (N.D.).

## REFERENCES

Acres, R. B., Conlon, P. J., Mochizuki, D. Y. and Gallis, B., 1986, Rapid phosphorylation and modulation of the T4 antigen on cloned helper T cells induced by phorbol myristate acetate or antigen, J. Biol. Chem. 261:16210-16214.

Allison, J. P. and Lanier, L. L., 1987, Structure, function, and serology of the T-cell antigen receptor complex, Annu. Rev. Immunol. 5:503-540.

Bangham, A. D., Standish, M. M. and Watkins, J. C., 1965, Diffusion of univalent ions across lamellae of swollen phospholipids, J. Mol. Biol. 13:238-252.

Carrel, S., Moretta, A., Pantaleo, G., Tambussi, G., Isler, P., Perussia, B. and Cerottini, J.-C., 1988, Stimulation and proliferation of CD+ peripheral blood T lymphocytes induced by an anti-CD4 monoclonal antibody, Eur. J. Immunol. 18:333-339.

Centers for Disease Control, 1988, 1988, Agent summary statement for human immunodeficiency virus and report on laboratory-acquired infection with human immunodeficiency virus, Morbid. Mortal. Week. Rep. (Suppl. No. S-4) 37:1-17.

Cheng-Mayer C., Rutka, J. T., Rosenblum, M. L., McHugh, T., Stites, D. P. and Levy, J. A., 1987, Human immunodeficiency virus can productively infect cultured human glial cells, Proc. Natl. Acad. Sci. U.S.A. 84:3526-3530.

Chesebro, B., Buller, R., Portis, J. and Wehrly, K., 1990, Failure of human immunodeficiency virus entry and infection in CD4-positive human brain and skin cells, J. Virol. 64:215-221.

Clapham, P. R., Weber, J. N., Whitby, D., McIntosh, K., Dalgleish, A. G., Maddon, P. J., Deen, K. C., Sweet, R. W. and Weiss, R. A., 1989, Soluble CD4 blocks the infectivity of diverse strains of HIV and SIV for T cells and monocytes but not for brain and muscle cells, Nature (London) 337:368-370.

Clapham, P. R., Weiss, R. A., Dalgleish, A. G., Exley, M., Whitby, D. and Hogg, N., 1987, Human immunodeficiency virus infection of monocytes and T-lymphocytic cells: receptor modulation and differentiation induced by phorbol esters, Virology 158:44-51.

Dalgleish, A. G., Beverly, P. C. L., Clapham, P. R., Crawford, D. H., Greaves, M. F. and Weiss, R. A., 1984, The CD4 (T4) antigen is an essential component of the receptor for the AIDS retrovirus, Nature (London) 312:763-767.

Dewhurst, S., Sakai, K., Bresser, J., Stevenson, M., Evinger-Hodges, M. J. and Volsky, D. J., 1987, Persistent productive infection of human glial cells by human immunodeficiency virus (HIV) and by infectious molecular clones of HIV, J. Virol. 61:3774-3782.

Doyle, C. and Strominger, J. L., 1987, Interaction between CD4 and class II MHC molecules mediates cell adhesion, Nature (London) 330:256-259.

Düzgüneş, N., Wilschut, J., Hong, K., Fraley, R., Perry, C., Friend, D. S., James, T. L. and Papahadjopoulos, D., 1983, Physicochemical characterization of large unilamellar vesicles prepared by reverse-phase evaporation, Biochim. Biophys. Acta 732:289-299.

Düzgüneş, N., Larsen, C. E., Konopka, K., Alford, D. R., Young, L. J. T., McGraw, T. P., Davis, B. R., Nir, S. and Jennings, M., 1991, Fusion of HIV-1 and $SIV_{mac}$ with liposomes and modulation of HIV-1 infectivity, in: "Mechanisms and Specificity of HIV Entry into Host Cells," N. Düzgünes, ed., Plenum Press, New York (this volume).

Emmerich, F., Kanz, L. and Eichman, K., 1987, Cross-linking of the T cell receptor complex with the subset-specific differentiation antigen stimulates interleukin 2 receptor expression in human CD4 and CD8 T cells, Eur. J. Immunol. 17:529-534.

Felgner, P. L., Gadek, T. R., Holm, M., Roman, R., Chan, H. W., Wenz, M., Northrop, J. P., Ringold, G. M., and Danielsen, M., 1987, Lipofectin: a highly efficient, lipid-mediated DNA-transfection procedure, Proc. Natl. Acad. Sci. U.S.A. 84:7413-7417.

Gartner, S., Markovitz, P., Markovitz, D. M., Kaplan, M. H., Gallo, R. C. and Popovic, M., 1986, The role of mononuclear phagocytes in HTLV-III/LAV infection, Science 233:215-219.

Gay, D., Maddon, P., Seklay, R., Talle, M. A., Godfrey, M., Long, E., Goldstein, G., Chess, L., Axel, R., Kappler, J. and Marrack, P., 1987, Functional interaction between human T cell protein CD4 and the major histocompatibility complex HLA-DR antigen, Nature (London) 328:626-629.

Habeshaw, J. A. and Dalgleish, A. G., 1989, The relevance of HIV env/CD4 interactions to the pathogenesis of acquired immune deficiency syndrome, J. AIDS 2:457-468.

Harouse, J. M., Kunsch C., Hartle, H. T., Laughlin, M. A., Hoxie, J. A., Wigdahl, B. and Gonzales-Scarano, F., 1989, CD4-independent infection of human neural cells by human immunodeficiency virus type 1, J. Virol. 63:2527-2533.

Ho, D. D., Rota, T. R. and Hirsch, M. S., 1986, Infection of monocyte/ macrophages by human T lymphotropic virus type III, J. Clin. Invest. 77:1712-1715.

Hoffman, A. D., Banapour, B. and Levy, J. A., 1985, Characterization of the AIDS-associated retrovirus reverse transcriptase and optimal conditions for its detection in virions, Virology 147:326-335.

Hope, M. J., Bally, M. B., Webb, G. and Cullis, P. R., 1985, Production of large unilamellar vesicles by a rapid extrusion procedure. Characterization of size distribution, trapped volume and ability to maintain a membrane potential, Biochim. Biophys. Acta 812:55-65.

Hoxie, J. A., Matthews, D. M., Callahan, K. J., Cassel, D. L. and Cooper, R. A., 1986, Transient modulation and internalization of T4 antigen induced by phorbol esters, J. Immunol. 137:1194-1201.

Janeway, C. A., Haque, S., Smith, L. A. and Saizawa, K., 1987, The role of the murine L3T4 molecule in T cell activation: differential effects of anti-L3T4 on activation by monoclonal anti-receptor antibodies. J. Mol. Cell. Immunol. 3: 121-131.

Klatzmann, D., Champagne, E., Chamaret, S., Gruest, J., Guetard, B., Hersend, T., Gluckman, J. C. and Montagnier, L., 1984, T-lymphocyte T4 molecule behaves as the receptor for human retrovirus LAV, Nature (London) 312:767-768.

Konopka, K., Davis, B. R., Larsen, C. E., Alford, D. R., Debs, R. J. and Düzgüneş, N., 1990a, Inhibition and enhancement of HIV-1 infectivity by liposomes, Abstr. VI Intl. Conf. AIDS (p.145), San Francisco, CA.

Konopka, K., Davis, B. R., Larsen, C. E., Alford, D. R., Debs, R. J. and Düzgüneş, N., 1990b, Liposomes modulate human immunodeficiency virus infectivity, J. Gen. Virol. 71:2899-2907.

Larsen, C. E., Alford, D. R., Young, L. J. T., McGraw, T. P. and Düzgüneş, N., 1990, Fusion of simian immunodeficiency virus ($SIV_{mac}$) with liposomes and erythrocyte ghosts: effect of liposome composition, low pH and calcium, J. Gen. Virol. 71:1947-1955.

Li, X. L., Moudgil, T., Vinters, H. V. and Ho, D. D., 1990, CD4-independent, productive infection of a neuronal cell line by human immunodeficiency virus type 1, J. Virol. 64:1383-1387.

Maddon P. J., Dalgleish, A. G., McDougal, J. S., Clapham, P. R., Weiss, R. A. and Axel, R., 1986, The T4 gene encodes the AIDS virus receptor and is expressed in the immune system and the brain, Cell 47:333-348.

Maddon, P. J, Littman, D. R., Godfrey, M., Maddon, D. E., Chess, L. and Axel, R., 1985, The isolation and nucleotide sequence of a cDNA encoding the T cell surface protein T4: a new member of the immunoglobulin family, Cell 42: 93-104.

McDougal, S. J., Nicholson, J. K. A., Cross, G. D., Cort, S. P., Kennedy, M. S. and Mawle, A. C., 1986, Binding of the human retrovirus HTLV-III/LAV/ARV/HIV to the CD4 (T4) molecule: conformation dependence, epitope mapping, antibody inhibition, and potential for idiotypic mimicry, J. Immunol. 137:2937-2944.

Reinherz, E. L. and Schlossman, S. F., 1980, The differentiation and function of human T lymphocytes, Cell 19: 821-827.

Rivas, A., Takada, S., Koide, J., Sonderstrup-McDevitt, G. and Engleman, E. G., 1988, CD4 molecules are associated with the antigen receptor complex on activated but not resting cells, J. Immunol. 140:2912-2918.

Robey, E. and Axel, R., 1990, CD4: collaborator in immune recognition and HIV infection, Cell 60:697-700.

Saizawa, K. K., Rojo, J. and Janewey, C. A., 1987, Evidence for a physical association of CD4 and the CD3:a:b T-cell receptor, Nature (London) 328:260-263.

Sattentau, Q., Dalgleish, A. G., Weiss, R. A. and Beverly, P. C. L., 1986, Epitopes of the CD4 antigen and HIV infection, Science 234:1120-1123.

Sleckman, B.P., Peterson, A., Jones, W. K., Foran, J. A., Greenstein, J. L., Seed, B. and Burakoff, S., 1987, Expression and function of CD4 in a murine T-cell hybridoma, Nature (London) 328:351-353.

Sodroski, J., Goh, W. C., Rosen, C., Campbell, K. and Haseltine, W. A., 1986, Role of the HTLV/LAV envelope in syncytium formation and cytopathicity, Nature (London) 322:470-474.

Szoka, F., Olson, F., Heath, T., Vail, W., Mayhew, E. and Papahadjopoulos, D., 1980, Preparation of unilamellar liposomes of intermediate size (0.1-0.2 µm) by a combination of reverse phase evaporation and extrusion through polycarbonate membranes, Biochim. Biophys. Acta 601:559-571.

Tang, S. and Levy, J. A., 1990, Parameters involved in the cell fusion induced by HIV, AIDS 4:409-415.

Tateno, M., Gonzales-Scarano F. and Levy, J. A., 1989, Human immunodeficiency virus can infect CD4-negative human fibroblastoid cells, Proc. Natl. Acad. Sci. U.S.A. 86:4287-4290.

Weber, J., Clapham, P., McKeating, J., Stratton, M., Robey, E. and Weiss, R., 1989, Infection of brain cells by diverse human immunodeficiency virus isolates: role of CD4 as receptor, J. Gen. Virol. 70:2653-2660.

Weiss, R. A., Clapham, P. R., McClure M. and Marsh, M., 1989, The CD4 receptor for the AIDS virus, Biochem.Soc. Trans. 17:644-647.
Werner, A., Winskowsky, G., Cichutek, K., Norley, S. G. and Kurth R., 1990, Productive infection of both CD4+ and CD4- human cell lines with HIV-1, HIV-2 and $SIV_{agm}$, AIDS 4:537-544.

# THE ENDOCYTIC PATHWAY FOR HUMAN IMMUNODEFICIENCY VIRUS INFECTION

**C. David Pauza**

**Department of Pathology and Laboratory Medicine, University of Wisconsin Medical School, 1300 University Avenue Madison, WI 53706**

## Introduction

In this review, the pathway for infectious entry of HIV will be examined. Particular attention will be payed to analyzing the rates and efficiency of HIV binding, internalization, and uncoating. These kinetic parameters define the pathway for HIV penetration of T cells and monocytes and can be used to distinguish among the proposed routes for infectious entry of this virus. It is of practical significance to define the rate-limiting steps in the entry mechanism, as these reactions will be the most suitable targets for antiviral agents designed to interfere with HIV infection. These rate-limiting steps for infection of T cells and monocytes are compared in light of the kinetic analyses of virus entry discussed here.

Our data provide a comprehensive examination of HIV penetration of the T cell line CEM and the monocyte cell line U937. These systems represent the major target cell types for natural HIV infection. It is shown that HIV infection of T cells proceeds mainly via receptor-mediated endocytosis and that this is likely to be the pathway for infectious penetration of monocytes. In addition, virus enters the monocyte via a phagocytic pathway that is independent of specific HIV binding to the CD4 receptor. The latter mechanism, and the demonstration that this is a non-infectious route of entry (38), reveals features of the antiviral capacity of the monocyte in its role as an effector cell of natural immunity.

*Mechanisms and Specificity of HIV Entry into Host Cells*
Edited by N. Düzgüneş, Plenum Press, New York, 1991

The in vitro characterization of virus entry mechanisms provides a framework for evaluating HIV dissemination in vivo. Our understanding of the pathogenesis of AIDS depends ultimately on the precise characterization of the basic mechanisms of virus infection, replication, dissemination, and cytopatholgy. These studies represent an initial step towards evaluating the first component in the pathway leading to HIV infection and AIDS.

## Preparation and Characterization of Virus Stocks

Preparation of virus stocks: Virus stocks were prepared by acute infection of the lymphoid cell line CEM or from cell-free supernatants of the persistently infected cell line HCEM (37); the LAV1 isolate of HIV-1 (5) is our standard virus strain. The cell-free supernatants contained approximately 1 x $10^6$ tissue culture infectious doses per ml (TCID, determined by the ability to induce syncytium formation in the MT-2 cell line, ref. 13). Virus was concentrated by polyethylene glycol precipitation (39) or by centrifugation for 2 h. at 18,000 r.p.m. in a Beckman J2-21 centrifuge. The virus pellets were resuspended in 0.01 of the original culture volume in a buffer containing 50 mM NaCl, 5 mM EDTA, 10 mM Tris-HCl; pH 7.4. In general, PEG-precipitation recovers most efficiently the reverse transcriptase activity (> 85% yield) with some loss of infectivity (~ 45% yield). Direct centrifugation of cell-free supernatants (without PEG precipitation) is less efficient for recovery of reverse transcriptase activity (~ 65% yield) however, the relative infectivity of this material is higher (~ 85%). Accordingly, most experiments relating to the biochemical analysis of virus binding and entry were performed with virus prepared by direct centrifugation to ensure the highest possible biological activity.

Radiolabeled virus stocks were also prepared. Logarithmically growing CEM cells were infected with 1 TCID per cell and cultured for 3 days. At this point in the infection, greater than 50% of the cells were present as multinucleated syncytia. The cells were recovered by centrifugation and resuspended in an appropriate medium. For $^{32}P$ labeling of HIV, the infected cells were resuspended in DME lacking phosphate (GIBCO Laboratories, Grand Island, New York); the medium was supplemented with 10% dialyzed, fetal bovine serum and the cell concentration was adjusted to 5 x $10^5$ per ml. Carrier-free $H_3{}^{32}PO_4$ (New England Nuclear, Boston, MA) was added to a final concentration of 1 mCi per ml and the labeling proceeded for 5 h. At the end of this period, the cells were removed and the supernatant virus was concentrated by centrifugation. The virus was purified further by centrifugation through a sucrose step-gradient in which the upper phase was 22% sucrose and the lower phase was 50% sucrose (21); radiolabeled virus was collected from the interface after 16 h of centrifugation at 16,500 r.p.m. in a Beckman J2-21 centrifuge. A typical labeling experiment yielded 1 ml of radiolabeled virus stock containing $10^7$ Cerenkov counts per ml and approximately 1 x $10^5$ TCID. Based on the physical to infectious particle ratio (discussed below) the specific activity of these $^{32}P$-labeled virus preparations was approximately 0.5 cpm Cerenkov per particle.

Virus stocks containing $^3H$-leucine were prepared. CEM cells were infected as described above. On day 3 post-infection, the cells were collected by centrifugation, resuspended in DME lacking leucine and supplemented with 10% dialyzed, fetal bovine serum. Then, L-[4,5-$^3H$(N)]-leucine (specific activity 52.3 Ci/mmole) was added at 0.25 mCi per ml of culture. After an overnight incubation, the cells were removed by centrifugation and the virus was collected and purified as described for the $^{32}P$-labeled virus stocks. A typical experiment yielded 2 ml of virus stock with a titer of 1 x $10^7$ TCID/ml, and a specific activity of 200 cpm per µl.

<u>Characterization of labeled virus preparations</u>: $^{32}P$-labeled virus preparations were characterized by two assays. First, the total incorporation into virions was determined by immunoprecipitation with serum antibodies from HIV-seropositive individuals. Second, the virus preparations were disrupted by treatment with NP-40 detergent ; RNaseA was added to a final concentration of 25 µg/ml and the samples were incubated at 37° for 30 minutes. At the end of this nuclease treatment, the samples were precipitated with 10% cold TCA. The TCA-precipitable counts were determined and compared to the counts for samples not treated with NP-40. In most preparations, 90% of the TCA-precipitable counts were eliminated by prior treatment of the virions with NP-40. The fact that detergent disruption was required to

render the $^{32}P$-labeled material susceptible to RNaseA treatment and the observation that a large percentage of the detergent-solubilized, radioactive material was degraded by RNaseA treatment, indicated that >90% of the label was incorporated into viral RNA. Phosphoproteins and phospholipids did not appear to make up a significant percentage of the labeled species.

$^{3}H$-labeled virus preparations were characterized in three ways. First, the total counts were determined by immunoprecipitation with patient antisera. Second, the virus titer was determined by the syncytium formation assay (13) and compared to unlabeled virus stocks in order to ensure that the ratio of enzyme activity to infectivity, was the same for both preparations. Third, the specific radioactivity of the virus preparation was determined. Total cellular protein content was measured by the Bradford dye binding assay (Biorad). Specific activity of this material was measured by liquid scintillation counting. In three independent preparations, the specific radioactivity of $^{3}H$-leucine labeled HIV averaged 0.02 cpm/TCID.

Determination of the particle to infectious unit ratio: The $^{3}H$-leucine labeled HIV preparations were used to measure the approximate number of virus particles contained within a single infectious unit. As noted above, we utilized the syncytium-formation assay described by Haertle (13) to determine the infectivity of each preparation; the average infectivity value for our preparations was 1 x $10^{7}$TCID per ml. We have assumed that the particle mass of HIV is similar to that reported previously for murine retroviruses. According to Andersen and Nexo (3), the particle is $10^{9}$g/mole and 8% leucine by weight. The specific activity of the labeled viral proteins was assumed to be the same as the specific activity of total cellular protein and is equivalent to 37,500 cpm/µg; the specific activity of the label was approximately 55 Ci/mmole. On the basis of these values, we calculate that the virus preparation contains 2.18 x $10^{6}$ particles/µl. Considering that the infectivity value for these materials was $10^{4}$ TCID/µl, it is possible to calculate that there are approximately 200 viral particles per TCID. This value is in good agreement with the value of 0.5-1.0% infectious particles reported previously for murine retroviruses (3). It is important to note however, that these calculations do not distinguish between the situation where 1 in 200 particles is infectious and the case in which simultaneous infection by 200 particles is required. This distinction may be of some significance for our understanding of the productive route of viral infection.

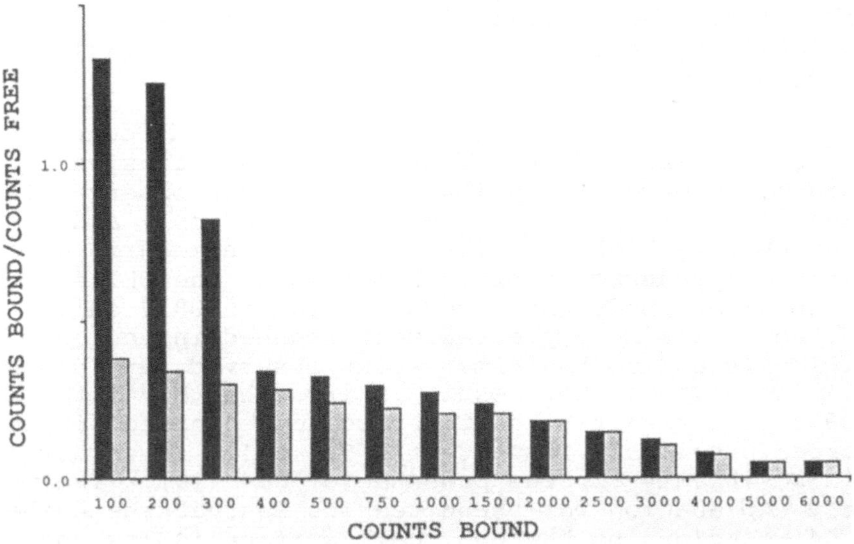


Figure 1. Binding of $^{32}$P-labeled HIV to CEM (solid bars) or U937 cells (stippled bars).

Binding of $^{32}$P-labeled virus to T cells and monocytes: Two cell lines representing the T lymphocyte and monocyte lineages have been used in all of these studies. The T lymphocyte line is CEM and the monoblastoid cell line is U937 (43).

Standard conditions for quantitative binding of virus to cells were: $10^3$ cells/ml and volumes of virus containing 2 x $10^3$ to 5 x $10^5$ cpm/sample of $^{32}$P-labeled HIV. This represents 4 to 1,000 viral particles per cell. The higher particles to cell ratios were required to approach saturation binding. The virus plus cell mixtures were incubated at 4° for 1 h and the unbound particles were removed by spinning the cells through a 0.5 ml cushion of undiluted bovine serum. The use of high protein content buffer solutions to wash cells reduced background and decreased the variability among experiments. The technique was developed initially for quantitative analysis of transferrin binding to cell membranes (36).

The results of binding assays are shown in Fig. 1. The data demonstrate abundant high affinity binding of HIV to the T cell line CEM. Some low affinity binding was also observed at higher input virus concentrations. The high affinity binding was blocked efficiently by pretreatment of the cells with 5 µg/ml of OKT4A monoclonal antibody that masks specifically the HIV binding site of the CD4 receptor (22).

A different pattern of virus binding was detected for the monoblastoid cell line U937. The virus:cell interactions were of the low affinity type observed for CEM cells, at high input virus concentrations, and there was no obvious high affinity binding. The degree of inhibition produced by pretreatment with the OKT4A monoclonal antibody was also decreased in U937 cells, even though antibody pretreatment reduced infection efficiently and to the same extent observed for the CEM cell line. The latter result affirms that HIV infection of U937 cell requires the CD4 receptor. Indeed, it was reported previously that 70% of U937 cells in culture stain positively for the presence of CD4 (10). The most likely explanation for these results is that the number of CD4 receptors on the U937 cell surface is low and the specific activity of radiolabeled virus was insufficient to detect accurately the high affinity interactions. However, these data cannot rule out the possibility that the binding of HIV to U937 cells is qualitatively different from virus binding to T cells.

## Biochemical Analysis of HIV Entry Into T cells and Monocytes

Method for biochemical analysis of HIV penetration: The pathway for HIV uptake and uncoating was studied using $^{32}$P-radiolabeled HIV. Advantage was taken of the fact that >95% of the radiolabel incorporated into virions was contained within viral RNA and that this material was resistant to nucleolytic degradation as long as it was contained within an intact viral particle. Once the RNA was released from the particle, it was degraded readily by RNaseA. Accordingly, the $^{32}$P-labeled HIV preparations provided a suitable substrate for the analysis of individual steps in the viral penetration mechanism.

Radiolabeled virions were preincubated with cells at 4° and internalization was initiated synchronously by warming the mixtures to 37°. At various intervals post-warming, the subcellular distribution of HIV into four subsets was evaluated (38): 1) Cells were washed by spinning through a serum cushion; the radioactivity remaining was defined as the total *cell-associated* counts. 2) Cells were treated briefly with trypsin (to remove material present only at the cell surface) and then washed; this fraction represented the *internalized* virus. 3) The cells were lysed, by three cycles of freeze-thawing, and the macromolecular RNA present in this lysate was recovered by TCA precipitation. The TCA-precipitable RNA was defined as the *coated plus uncoated* RNA fraction to indicate that the material was present either within viral particles (coated) or free in the cytoplasm in a high molecular weight form (uncoated). 4) Finally, the cell lysates were treated with RNaseA, to degrade any viral RNA present as uncoated nucleic acid; the remaining TCA-precipitable counts defined the *coated* RNA, to indicate that its biochemical properties were consistent with viral RNA still present within the virion. This approach was fashioned after the method of Gollins and Porterfield (12) used previously to study the pathway for infectious entry of flavivirus particles.

<u>HIV penetration of CEM cells</u>: The results of our analyses of HIV penetration in both CEM and U937 cells are shown in Fig. 2. Virus internalization is relatively slow and inefficient in the T cell line; the half-time for internalization is approximately 15 minutes and no more than 60% of the cell associated material ever appears in the trypsin-resistant fraction. Once internalized, the viral RNA remains relatively intact; this conclusion was reached by comparing the curve for internalized RNA (filled boxes) to the coated plus uncoated RNA curve (filled triangles). These curves are fairly similar thus, indicating that viral RNA present within the trypsin-resistant fraction remains greater than 80% TCA-precipitable. Accordingly, the rates of RNA degradation, due either to nucleolysis of uncoated RNA or digestion of virus particles, are minimal in the CEM line. The decrease in coated RNA (filled circles) reveals the rate of RNA uncoating. As the amount of internalized, nuclease-resistant RNA decreases, one can infer that the amount of uncoated RNA increases. Therefore, the rate of viral uncoating is derived roughly from the inverse of the coated RNA curve. The half-time for this process is approximately

75 minutes and greater than 80% of the internalized HIV will eventually uncoat and release intact viral RNA into the cytoplasm. Virus penetration is slow and inefficient in CEM cells, even though the internalized material is uncoated efficiently and remains stable within the cytoplasm for extended periods.

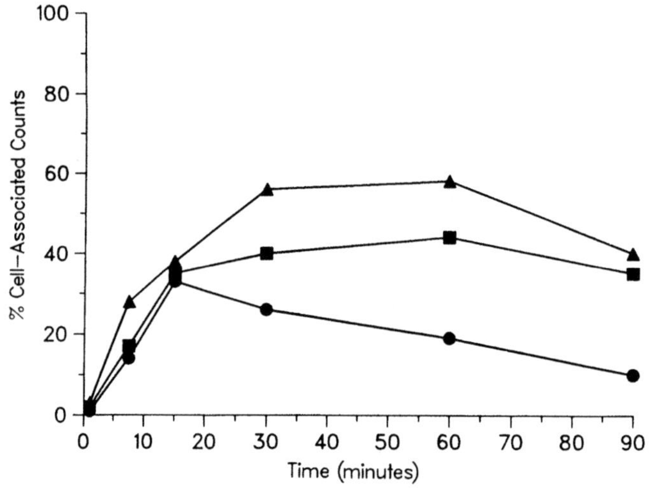


Figure 2. Kinetic analysis of $^{32}P$-labeled HIV internalization in CEM cells. Filled triangles represent internalized virions. Filled boxes denote internalized, coated plus uncoated RNA. The filled circles indicate that fraction of internalized viral RNA remaining resistant to ribonuclease digestion.

Biochemical analysis of HIV entry into CEM cells revealed that a significant proportion of cell-associated virus was present as internalized, intact viral particles. This conclusion was based on the direct observation of trypsin-resistant (hence internal), RNaseA-resistant (hence intact) virions. Therefore, these data demonstrate that HIV enters T cells via endocytosis and that this pathway for virus penetration accounts for the majority of HIV particles interacting with these cells. Furthermore, these data do not support a role for direct fusion between the viral and plasma membranes. If the direct fusion pathway were to represent a significant fraction of the infecting virus particles, it should have been possible to observe internalized, uncoated viral RNA prior to the appearance of internalized virions. Such a finding was not made and it is safe to conclude that fusion between the virion and plasma membrane occurs at a very low frequency.

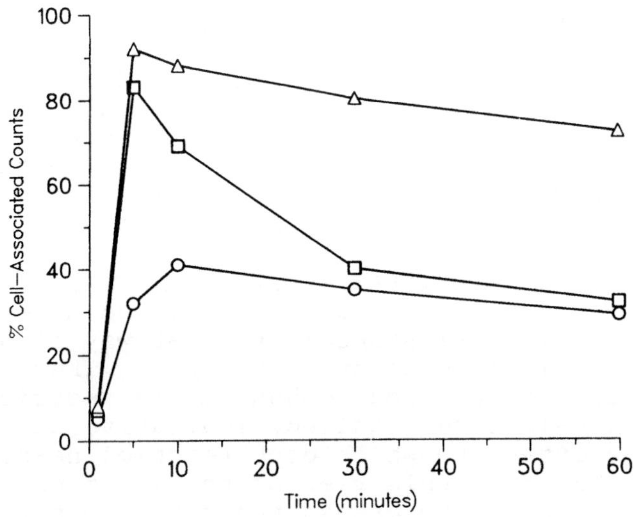


Figure 3. Kinetic analysis of $^{32}P$-labeled HIV internalization in U937 cells. Triangles represent internalized virions. Boxes denote internalized, coated plus uncoated RNA. Circles indicate that fraction of internalized viral RNA remaining resistant to ribonuclease digestion.

<u>HIV penetration of U937 cells</u>: The pattern of HIV penetration in U937 cells is markedly different from that described above for T cells. In this monocyte cell line, virus internalization is extremely rapid and efficient. Greater than 90% of cell-associated counts become trypsin-resistant within 5 minutes post-warming. The internalized material follows a different pathway than the one observed previously for T cells. In figure 3, the coated plus uncoated RNA curve (open triangles) diverges sharply from the internalized curve (open

boxes). Rapid disappearance of TCA-precipitable material indicates that most of the internalized virus is degraded within a few minutes. Indeed, at 30 minutes post-warming, approximately 50% of the internalized viral RNA is no longer TCA-precipitable. The curve representing coated RNA (open circles) has an unusual shape and is not directly interpretable.

It is most likely that the rapid rates of internalization, and subsequent degradation, obscure the true kinetics of virus uncoating. Accordingly, it is not possible to derive from these data, a rate for HIV uncoating in U937 cells. Virus internalization is rapid and extremely efficient in U937 cells, despite the lack of abundant, high affinity CD4 receptors. Once internalized, the virus follows mainly a degradative pathway. This dominant pathway for internalization and degradation obscures slower rates of HIV uptake that may occur. We presume that virus fusion occurs in the endosome. However, these data do not provide definitive evidence about the mechanism of penetration.

The rate limiting steps for HIV infection: The data presented previously (38) and reviewed here, characterize individual steps in the pathway of HIV internalization in T cells and monocytes. These elements of the penetration mechanism are described in two ways: the half-time required, indicative of reaction rate, and the extent to which each reaction nears completion. Based on this view of the kinetics of HIV entry, it is possible to assess the rate-limiting step for viral infection in each of these cell types.

During HIV penetration of CEM cells, the slowest step in the pathway is virus uncoating; this reaction proceeds with a half-time of approximately 75 minutes. Even though this step has the lowest rate constant, it does not appear to be limiting for HIV infection of T cells. Events subsequent to HIV internalization may proceed more slowly however, they occur inevitably once the virus enters the endocytic pathway. Therefore, the initial virus internalization step seems most likely to be the rate-limiting reaction. Thus, the subsequent steps in this complex pathway are judged to be unavoidable consequences of the initial rate of virus internalization; only the internalization rate is limiting.

It is more difficult to assign a rate limiting step for HIV infection of U937 cells. In this cell line, internalization is a rapid and efficient process and does not appear to be rate limiting for infection. Here, the most significant feature of virus entry is the rapid degradation of incoming virus particles. Therefore, escape from degradation is crucial for successful HIV infection of U937 cells. Discrimination between degradative and infectious entry pathways is probably related to the mechanisms responsible for the initial events in HIV penetration. It seems likely that HIV enters U937 cells via both CD4-dependent and CD4-independent pathways and it is tempting to speculate that one mechanism leads to degradation of the incoming virus particle and the other leads to productive infection.

Infection of U937 cells is blocked efficiently by pretreatment with the OKT4A monoclonal antibody (4,9,38). Therefore, the small amount of CD4-dependent HIV binding to U937 cells characterizes the subpopulation of cell-associated virus capable of establishing a productive infection. The rate limiting step for HIV infection of U937 cells pertains to the fraction of HIV that is CD4-associated, versus the fraction bound to the cell via other interactions, and the efficiency of internalization leading to the phagocytic pathway. It seems most simple to state that the level of U937 cell phagocytic activity limits the extent of HIV infection. The rate limiting mechanism is the competition between the CD4-dependent and CD4-independent modes of HIV internalization reflecting indirectly, discrimination between productive entry and degradation in the phagosome.

**Ultrastructural analysis of HIV penetration mechanisms**

Electron microscopy was performed to provide corroborative evidence about the mechanism for HIV penetration of susceptible cells. Innocula were prepared and concentrated as described previously (38). These high titer virus preparations were added to CEM or U937 cells at 4°; virus uptake and uncoating was initiated synchronously by warming the cells to 37°. Samples were taken at short intervals, up to 30 minutes post-warming. The cells were washed and resuspended in cold phosphate buffered saline. Immediately, the cells were fixed by glutaraldehyde and osmium tetroxide according to the method of Hirsch and Fedorko (16). After dehydration through a graded series of ethanol solutions and embedding in epoxy resin, the samples were sectioned and stained with saturated methanolic uranyl acetate and lead citrate prior to electron microscopy (38).

The infections performed for electron microscopy utilized approximately 30 TCID per cell. Based on the value of 200 particles per TCID, this means that approximately 6,000 virus particles were added per cell. Although this is an unusually high multiplicity of infection, these conditions were required in order to produce electron micrographs revealing sufficient viral particles for detailed analysis. As will be shown later, this level of input virus did not affect significantly the pathway for HIV entry into CEM and U937 cells.

Ultrastructural analysis of HIV infection of CEM cells: The pathway for HIV infection of the CEM cell line is demonstrated in Fig. 4. This montage of electron micrographs represents the discrete stages in the viral entry mechanism at various points in time. Panel A shows the situation of HIV bound to the cell surface prior to warming. Here, numerous particles are associated with the cell surface. We examined in detail, more than 5,000 of these cell-associated particles; in no case did we observe evidence for direct fusion between the viral envelope and the plasma membrane of the cell. At 2.5 minutes post-warming, shown in Panel B, HIV particles are observed within invaginations of the plasma membrane. After 5 minutes at 37°, Panel C, the invaginations have resolved into fully formed endosomes and the initial appearance of HIV particles within these structures is observed. Panel D presents a similar picture at 10 minutes post-warming. Within 15 minutes after the temperature shift, a significant fraction of the total number of virus particles are observed within endosomes and, Panel F, these vesicles appear to move away from the cell surface toward the nucleus. Also shown in Panel F (arrowhead) is a structure apparently reflecting the initial stages of fusion between the viral and endosomal membranes. Note that Panel F depicts the situation at 20 minutes post-warming.

In order to characterize more completely interactions between HIV envelope and the endosomal membrane, we utilized the goniometer tilting stage. This device permits observation of the sample through 40° tilting and reveals some aspects of these structures in three dimensions. Examination of numerous examples

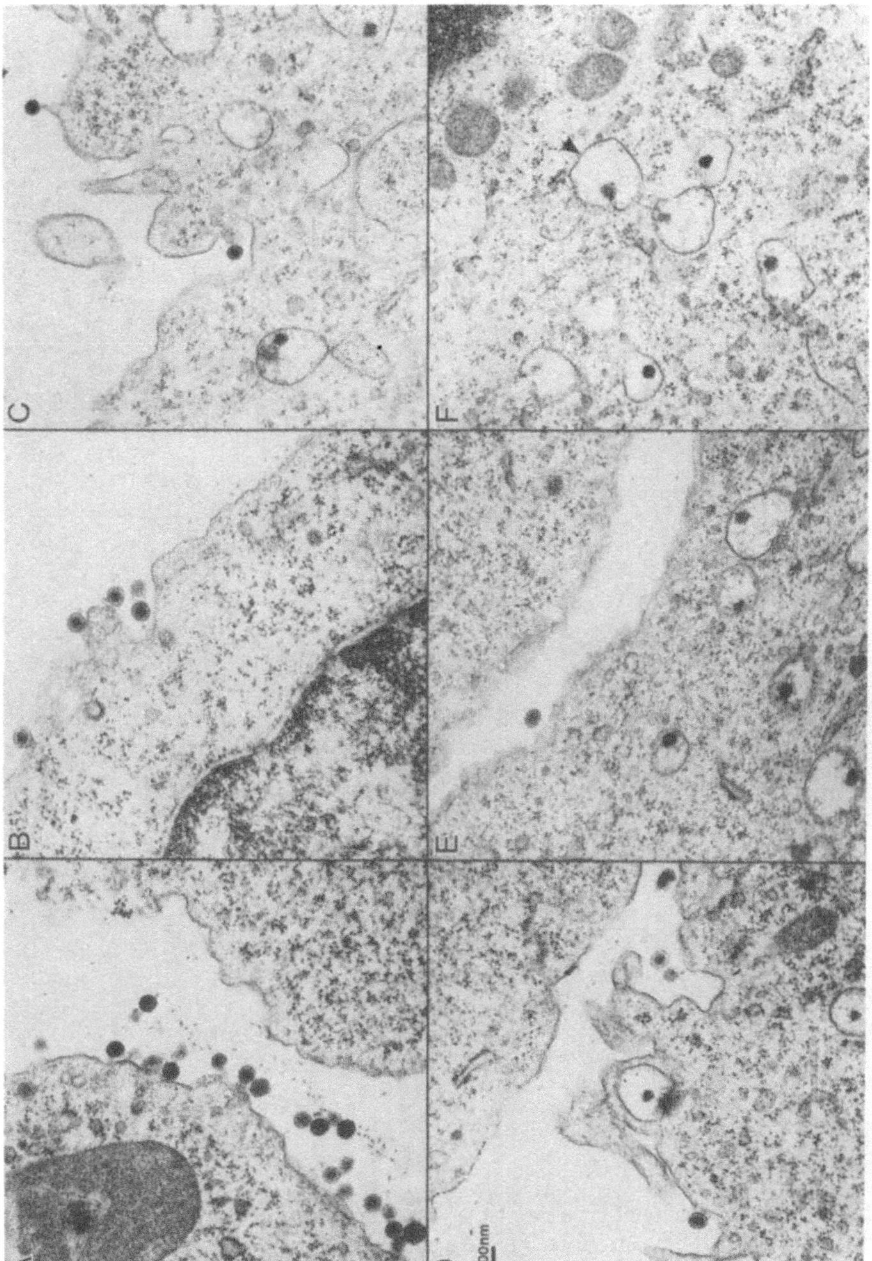


Figure 4. Ultrastructural analysis of the pathway for HIV penetration of CEM cells. The individual panels represent discrete time intervals throughout the initial 20 minutes of virus entry. Reprinted by permission of Rockerfeller University Press.

of virion:endosome interactions revealed a significant proportion of verifiable fusion events. In addition, a number of forms were observed where the disrupted virus particle was intercalated fully in the endosomal membrane (38). Structures of this type are consistent with late stages of the virion:endosome fusion mechanism (14). An enlarged view of one of these fusion events is shown in Figure 5.

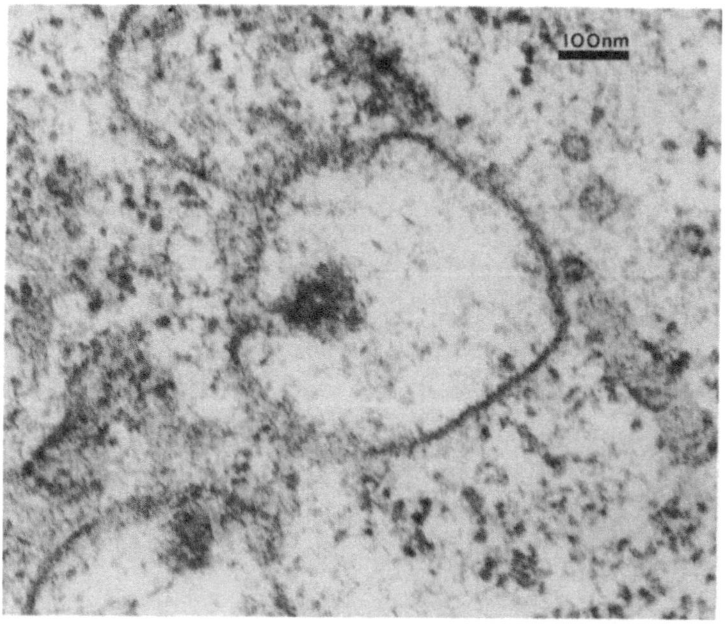


Figure 5. Magnified view of an HIV virion fusing with an endosomal membrane in CEM cells. This enlarged view was from Panel F of Figure 4.

These data corroborate the biochemical observations and demonstrate conclusively that HIV particles are present within endosomes and, in fact, uncoat via fusion between the viral and endosomal membranes. In order to assess the validity of these observations, we counted the frequency of virus particles that were cell surface-associated, present within endosomes, or undergoing intermembrane fusion. For each time point, at least 500 virus particles were observed; the data are summarized in Table 1. An important feature of the ultrastructural analysis of HIV entry is that this technique does not reveal the presence of uncoated viral RNA. Accordingly, the relative distribution of various forms cannot account for particles that have completed the pathway and only reveals the presence of those that are still in transit. This accounts for the relative infrequency of fusion figures (a maximum of 8% at 20 minutes post-warming) as compared to the the value of 30% uncoated RNA that can be derived from the biochemical data concerning HIV penetration of CEM cells.

**Table 1**

**Distribution of HIV Particles Post-Infection.**

| Time at 37° (minutes) | 0 | 2.5 | 5 | 10 | 15 | 20 | 60 |
|---|---|---|---|---|---|---|---|
| Cell surface | 100 | 96 | 92 | 74 | 62 | 47 | 35 |
| Within endosomes | 0 | 4 | 8 | 23 | 35 | 45 | 58 |
| Fusion figures | 0 | 0 | 0 | 3 | 3 | 8 | 7 |

(Values represent the percentage of particles identified at discrete positions within individual CEM cells. More than 500 individual virions were accounted for at each time point.)

Two features of this quantitative analysis of HIV entry are significant. First, these data are in good agreement with the biochemical observations presented previously. This encourages the view that the high multiplicity of infection utilized for preparation of electron microscopy samples, did not affect significantly the rates of virus internalization and uncoating. The second important point concerns the length of residency for HIV particles in endosomes. A significant number of virus particles were observed within endosomes even at 1 hour post-infection. Recalling the previous observation (Fig. 2) that internalization of HIV particles by CEM cells reaches a maximum within approximately 30 minutes post-infection, then it seems reasonable to conclude that these virus particles remain within endosomes for significant periods of time. This residence time has direct implications for the interpretation of data regarding the effects of lysosomotropic weak bases on the infectious entry of HIV (29,42). These agents increase the pH within acidic endosomes, and prevent virus uncoating in some systems (15,45), by decreasing the rate of fusion between the viral and endosomal membranes. Accordingly, the efficacy of these agents depends on two factors: the magnitude of the effect of altered pH on fusion rate and the length of time that the viral particle resides within the endosome. Should the effect of pH on HIV membrane fusion prove small, the prolonged residence of this virus in endosomes will provide ample opportunity for uncoating even in the presence of lysosomotropic weak bases. Accordingly, the reported inability of lysosomotropic weak bases to block HIV penetration, must be viewed cautiously until additional information becomes available regarding the rates of virion:membrane fusion.

Ultrastructural analysis of HIV infection of U937 cells: Much less is known about the mechanism for HIV penetration of U937 cells. The typical pattern of viral particles is shown in Fig. 6. Here, we observe large phagocytic vacuoles containing numerous viral particles. Note the morphological distinction between virus particles in the endosome of CEM cells (Fig. 4) and the particles present in the phagosomes of U937 cells. Virions in phagosomes are structurally intact with well defined, limiting membranes and distinct cores. This is contrasted to the virions with badly defined limiting membranes and diffuse cores, that were observed in the T cell endosomes. These differences in structure appear to be consistent with the fates of virions contained within these distinct vesicles. The endosomal virions in T cells are part of the infectious pathway for virus entry as opposed to the virions in phagosomes that are headed for a degradative compartment.

Convincing examples of virus particles within endosomes in U937 cells, were not observed. The micrographs were scanned extensively for evidence of single virions within vesicles and for the diffuse particle structure characteristic of viral particles in CEM cells. Although several candidate examples were identified, these did not provide sufficient evidence to conclude that they represent a significant fraction of the internalized HIV. This conclusion is consistent with the biochemical observations on the pathway of virus internalization in U937 cells. The rapid, CD4-independent, pathway for internalization dominates the kinetic analysis of virus uptake and seems also to represent the majority of structures that can be observed in the electron microscope.

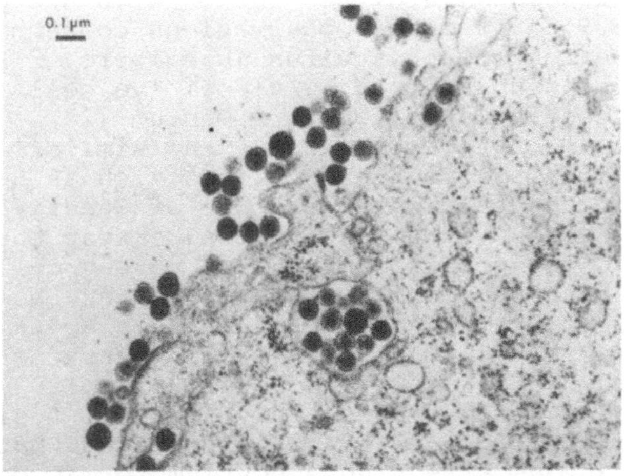


Figure 6. Electron micrograph of HIV virions within a U937 cell phagosome. Reprinted by permission of Rockefeller University Press.

## Discussion

The biochemical and ultrastructural analyses discussed here, support the concept that HIV penetration of susceptible cells occurs via receptor-mediated endocytosis. This is shown most convincingly for T cells, where it is clear that the majority of cell associated virus particles are internalized via the endosomal pathway. In addition, direct observation of fusion between viral and endosomal membranes establishes that this route of virus entry can lead to productive infection.

The more difficult problem concerns whether receptor-mediated endocytosis represents the only pathway for HIV entry into T cells and monocytes. Three types of data have been presented to suggest that direct fusion with the plasma membrane is, in fact, the biologically significant route of infection. The first type of argument concerns the capacity for HIV-infected cells to form multinucleated syncytia in culture. The second concern rests on the inability of lysosomotropic weak bases to inhibit HIV infection. The third area relates to the conclusions derived from mutagenesis studies on the CD4 receptor for HIV. Each of these arguments will be discussed briefly.

Initial support for the hyopothesis that HIV infection proceeds via direct fusion with the plasma membrane derives from the observations concerning the cytopathic effect of this virus in culture. Infected cells fuse with uninfected, CD4-positive cells in the neutral pH conditions of culture medium (25). This cell:cell fusion property recalls the similar characteristics of fusogenic viruses such as Sendai (30,34). In the particular example of Sendai, infection probably proceeds via direct fusion between the viral envelope and the plasma membrane of target cells (30) and it is thus, reasonable to surmise that the virus entry and cell:cell fusogenic mechanisms are similar. However, several studies initiated to characterize the cell:cell fusogenic activity of the HIV gp120 envelope protein indicate that virus entry and cell fusion are not necessarily linked. It was discovered that some but not all mutations affecting syncytium formation also reduced HIV infectivity (24,31,47). Accordingly, these analyses of the fusogenic potential of HIV distinguished clearly between glycoprotein functions contributing to infection and those responsible for syncytium formation. This distinction seems consistent with the known properties of HIV. Inactivated HIV particles are incapable of initiating cell:cell fusion. Therefore, the viral envelope of HIV induces cell:cell fusion only when expressed at high levels on the surface of infected cells. In this regard, the fusogenic potential of HIV is indistinguishable from that of vaccinia (20), Semliki Forest virus, influenza, or vesicular stomatitis virus (7,33,46), all of which are recognized as entering cells normally via receptor-mediated endocytosis. Accordingly, the cell:cell fusogenic activity of HIV may be involved in the mechanism of virus penetration, though the observation of syncytium formation does not, in itself, constitute a compelling argument about the pathway for HIV infection.

The second controversial feature of HIV infection concerns the pH dependence of virus penetration. Despite an initial report that inhibitors of endosomal acidification prevented HIV entry (26), it was reported subsequently by several laboratories (29,42), that infectious entry of this virus is resistant to the action of lysosomotropic agents; these authors concluded that HIV infection proceeds via direct fusion between the virion and the plasma membrane of infected cells. In cases where lysosomotropic weak bases inhibit infection effectively (reviewed in Marsh and Helenius, 28) it is safe to conclude that entry of the virus requires transport to, and acidification within the endosome. It seems less reliable to conclude that endocytosis is not the route for entry of HIV based on an inability to inhibit infection by lysosomotropic weak bases (29,42). Numerous examples of virions fusing to the endosomal membrane were presented here. In addition, the biochemical data measure directly the kinetics of virion uncoating within the endosome. Accordingly, it is reasonable to conclude that infection can occur via the endocytic pathway. So far, we have been unable to observe any evidence for direct fusion between the viral and plasma membrane. Thus, the main pathway for HIV internalization by T cells involves endocytosis.

Lysosomotropic weak bases reduce the rate of fusion between viral and endosomal membranes, by maintaining the pH of the vesicle within a range unfavorable to virus uncoating. Therefore, the effectiveness of these antiviral agents depends critically on: 1) the rate of virion:endosomal membrane fusion, 2) the pH-dependence of this rate, and 3) the length of time that the virus particle resides within the endosome. The chemical characterization of HIV envelope-mediated fusion events is underway (N. Düzgünes, this volume) and the preliminary observations suggest that this reaction is promoted in environments below neutral pH. In preliminary experiments in this laboratory, we have observed a decrease in the rate of virus uncoating in the presence of $NH_4Cl$ and amantidine, even though these was no detectable loss of infectivity. It seems likely that the rate of virion:endosome fusion is decreased in consequence of endosome neutralization however, the prolonged stay of HIV particles in these vesicles nevertheless, permits efficient infection. The kinetics of virion passage through this subcellular compartment are sufficiently slow to permit infection, even when the rate of intermembrane fusion is reduced by the action of lysosomotropic weak bases.

The third type of data opposing the thesis that HIV infection proceeds via receptor-mediated endocytosis, relies on mutational analysis of the CD4 receptor. It was shown previously that CD4 levels are modulated on the lymphocyte cell surface (18). In particular, the cell surface molecules are internalized rapidly subsequent to antigenic and mitogenic (44) stimulation. The most efficient means for modulating the surface levels of CD4 is treatment with phorbol ester compounds (9,19). The mechanism of phorbol ester-induced CD4 modulation has been explored in some detail.

Treatment of T cell lines with the phorbol ester tetradecanoyl phorbol acetate (TPA) induces an immediate phosphorylation of serine residues on the cytoplasmic tail of the CD4 molecule (6,18). Following phosphorylation, these molecules are removed rapidly from the cell surface (18). The biological significance of these phosphorylation-modulation observations is not well established. Cell surface CD4 is also modulated after antigenic stimulation of T cells (44) even though this treatment was reported not to induce receptor phosphorylation (18). Apparently, some of the normal biological functions of lymphocyte CD4 continue in the absence of receptor phosphorylation. Controversy also attends the issue of whether HIV binding induces either receptor phosphorylation or internalization. Hoxie, et al. (18) reported that purified gp120 bound to cells with high affinity although it was incapable of inducing CD4 phosphorylation. In contrast, Fields, et al. (11) demonstrated that binding of whole virus particles, instead of the monovalent, purified glycoprotein, induced CD4 phosphorylation in a time-dependent fashion. It is interesting to note that the maximum levels of CD4 phosphorylation occurred after 30 minutes exposure of the cells to HIV (11); this value agrees with our previous demonstration (38) that the maximum levels of T cell-associated HIV were observed within 30.

Mutagenesis of the CD4 receptor was employed to assess the functional significance of cytoplasmic serine residues in TPA-induced receptor internalization and HIV infection. Bedinger, et al. (6) engineered mutants of the CD4 molecule lacking cytoplasmic serine residues and, in some cases, lacking the entire cytoplasmic tail of CD4. These mutant receptor molecules were reintroduced into CD4(-) cell T cell lines and their properties were examined. The mutant receptor molecules, lacking the cytoplasmic tail (-cyt 401), were expressed more abundantly than the normal receptor molecule as evidenced by the increased relative fluorescence of the transfected cell lines.

Internalization of the -cyt 401 mutant was not induced by phorbol ester treament. Therefore, the cytoplasmic tail of CD4 appears to be required for phorbol ester-induced modulation of this cell surface molecule.

Deletion of the cytoplasmic tail of CD4 reduced significantly the susceptibiltiy to HIV infection. However, questions arise about the mechanism for virus entry in these mutant cell lines. The elegant structure-function analyses of CD4 provide substantial insight into one means for internalizing this cell surface molecule. One can argue , based on the previous studies by Hoxie, et al. (18) and Weyand, et al. (44), that phosphorylation of CD4 is not correlated with the normal biological functions of this protein. In addition, it is difficult to reconcile the relationship between surface density of the transfected protein, which appears to be unusually high in the cell lines transfected with the -cyt 401 mutant, and the efficiency of infection. Substantial levels of HIV replication were detected intially on day 5 post-infection for the wild-type receptor and on day 7 for the mutant receptor. Accordingly, the efficiency of infection was reduced substantially in the -cyt 401 mutant despite the presence of elevated numbers of receptors per cell surface.

## Summary

We have presented data supporting the view that HIV infection of T cells and monocytes proceeds via receptor-mediated endocytosis. Biochemical analysis of the pathway for internalization of $^{32}P$-labeled HIV revealed the presence of intact, intercellular virions and provided evidence for subsequent uncoating of these particles and release of viral RNA to the cytoplasm. Electron microscopy demonstrated virions within vesicles and documented fusion between viral and endosomal membranes.

Others have provided evidence supporting the notion that endocytosis is not required for HIV infection. In consequence of this apparent paradox, it has been suggested that both mechanisms, i.e. direct fusion to the plasma membrane and endocytosis, are functional pathways for HIV entry. I do not favor this position. Examination of the biology of CD4, its role in lymphocyte activation, and its function as the receptor for HIV imply a more active role for this molecule in virus infection than simply as a transient anchor for fusion events. The principal experiment of concern is the report by Maddon, et al. (26), that human CD4-expressing murine cell lines remain resistant to infection by VSV pseudotypes bearing the HIV envelope glycoprotein. This paper and a subsequent review of HIV entry by Marsh and Dalgleish (27) comment on the possibility that endocytosis of human CD4 may not occur in murine cells hence, the block to infectious entry of these pseudotype virions.

Resolution of the mechanism for HIV infection is more than rationalization or semantics. The pathway of endocytosis presents an unique feature of cell biology. Should HIV infection be proven to occur via this mechanism, then appropriate assays for novel classes of antiviral agents could be developed. Other features of the endosomal vesicle, such as the unique composition of its membrane (45) or the possible presence of specific proteolytic activities (2) might play important roles in HIV infection and thus, present new targets for the development of antiviral agents. Indeed, the synthetic liposome AL721 is an effective inhibitor of HIV infection in vitro (41). This antiviral agent probably acts by decreasing the cholesterol content of the virion envelope (32). It was shown previously that the envelope is specifically enriched in cholesterol (1), a property not associated with the plasma membrane but clearly involved in Semliki Forest virus fusion within the endosome (45). This type of approach, in which unique properties of the virus internalization pathway are exploited to predict targets for antiviral drug development, requires precise understanding of the virus entry mechanism.

Our knowledge of HIV infection mechanisms is also crucial for three other important areas of AIDS pathogenesis. First, recent reports attest to the validity of antibody (17) or complement (40) mediated enhancement of HIV infection. In both cases, it is likely that virus entry occurs via endocytosis of the ternary complex of HIV, antibody or complement, and the appropriate cell surface receptor. Once in the endosome, chemical reactions favoring HIV uncoating would be similar for this mechanism and the T cell uncoating reaction. It is reasonable to expect that inhibitors of virus uncoating in the lymphocyte endosme would also be effective for antibody, or complement-mediated infection.

The second area of interest impacted directly by our understanding of virus entry is the characterization of CD4-independent HIV infection (8). Here, it seems that non-specific binding of HIV to the cell surface, followed by passive internalization of this bound material, provides a reasonable working hypothesis for the mechanism of entry. Inefficient delivery of virus to the endosome would constitute a slow, but nonetheless productive route of entry.

The third aspect of AIDS pathogenesis concerns the mechanisms for virus dissemination within the infected host. It was proposed (35) that the depletion of CD4(+) cells in the infected individual reflects the fact that helper T lymphocytes must contact MHC Class II-positive antigen presenting cells in order to receive their initial activation signal. The CD4 molecule is required in this interaction and this obligatory requirement may influence significantly the tropism of HIV in vivo. Considering that macrophages, which are authentic antigen presenting cells, are thought to comprise the reservoir for virus (23,35), then it is reasonable to propose that HIV dissemination occurs principally in consequence of this cell:cell interaction and, only secondarily, via mechanisms involving the interaction of soluble virus particles with cell surface CD4. Therefore, the contribution of infection mechanisms involving soluble virus particles, to AIDS pathogenesis remains to be established.

Examination of HIV penetration mechanisms provides an unique insight into the biological mechanisms determining cellular susceptibility to infection. The studies already completed in this area constitute a thorough examination of the mechanics of virus infection. Continuing interest in this field holds the promise of greater detail and precision in our view of HIV penetration. Determining conclusively the route for infectious virus entry and establishing the importance of this mechanism in the natural history of HIV infection are two significant challenges awaiting investigators in this field. With increasing comprehension of the basic biology of HIV, it is likely that safe and effective AIDS therapies will be forthcoming.

**Acknowledgements**

Electron microscopy was performed by Todd Price. We also thank Gilbert Keller for helpful suggestions on sample preparation and José Galindo for expert assistance in the preparation of virus stocks. This work was supported by NIH Grant AI24591. The manuscript is dedicated to the memory of Mae M. Hollis.

**References**

1. Aloia, R., et al., (1988) "Lipid composition and fluidity of the human immunodeficiency virus (HIV)", Proc. Natl. Acad. Sci. (USA) **85**:900-904.

2. Andersen, K. (1987) "Cleavage fragments of the retrovirus surface protein gp70 during virus entry", J. Gen. Virol. **68**:2193-2202.

3. Andersen, K. and Nexo, B. (1983) "Entry of murine retroviruses into mouse fibroblasts", Virology **125**:85-98.

4. Asjo, B., et al. (1987) "Susceptibility to infection by the human immunodeficiency virus (HIV) correlates with T4 expression in a parental monocytoid cell line and its subclones", Virology **157**:359-365.

5. Barre-Sinousi, F., et al. (1983) "Isolation of T-lymphotropic retrovirus from a patient at risk for acquired immunodeficiency syndrome (AIDS)", Science **220**:868-871.

6. Bedinger, P., et al. (1988) "Internalization of the human immunodeficiency virus does not require the cytoplasmic domain of CD4", Nature **334**:162-165.

7. Chang-Fournier, F., Chang, C. and Lafay, F. (1977) "Mechanism of polykaryote induction by Vesicular Stomatitis virus in rat XC cells", J. Gen. Virol. **34**:305-314.

8. Clapham, P., et al. (1989) "Soluble CD4 blocks the infectivity of diverse strains of HIV and SIV for T cells and monocytes but not for brain and muscle cells", Nature **337**:368-370.

9. Clapham, P., et al. (1987) "Human immunodeficiency virus infection of monocytic and T lymphocytic cells: Receptor modulation and differentiation induced by phorbol ester", Virology **158**:44-51.

10. Dalgleish, A., et al. (1984) "The CD4 (T4) antigen is an essential component of the receptor for AIDS retrovirus", Nature **312**:763-767.

11. Fields, A., et al. (1988) "Human immunodeficiency virus induces phosphorylation of its cell surface receptor", Nature **333**:278-280.

12. Gollins, S. and Porterfield, J. (1984) "Flavivirus infection enhancement in macrophages: Radioactive and biological studies of antibody on viral fate", J. Gen. Virol. **65**:1261-1272.

13. Haertle, T., et al. (1988) "Metabolism and anti-HIV activity of 2'-halo-2',3'-dideoxyadenosine derivatives", J. Biol. Chem. **263**:2870-2875.

14. Helenius, A. (1984) "Semliki Forest virus penetration from endosomes: A morphological study", Biol. Cell **51**:181-186.

15. Helenius, A., Marsh, M. and White, J. (1982) "Inhibition of Semliki Forest virus penetration by lysosomotropic weak bases", J. Gen. Virol. **58**:47-61.

16. Hirsch, J. and Fedorko, M. (1968) "Ultrastructure of human leukocytes after simultaneous fixation with glutaraldehyde and osmium tetroxide and "post-fixation" in uranyl acetate", J. Cell Biol. **38**:615-627.

17. Homsy, J., Tateno, M. and Levy, J. (1988) "Antibody-enhancement of HIV infection", Lancet **i**:1285-1286.

18. Hoxie, J., et al. (1988) "T4 endocytosis and phosphorylation induced by phorbol esters but not by mitogen or HIV infection", J. Immunol. **140**:786-795.

19. Hoxie, J., et al., (1986) "Transient modulation and internalization of T4 antigen by phorbol esters", J. Immunol. **137**:1194-1201.

20. Janeczko, R., Rodriguez, J. and Esteban, M. (1987) "Studies on the mechanism of entry of Vaccinia virus in animal cells", Arch. Virol. **92**:135-150.

21. Kennedy-Stoskopf, S. and Narayan, O. (1986) "Neutralizing antibodies to Visna lentivirus: Mechanism of action and possible role in virus persistence", J. Virol. **59**:37-44.

22. Klatzmann, D., et al. (1984) "T-lymphocyte T4 molecule behaves as the receptor for human retrovirus LAV", Nature **312**:767-768.

23. Klatzmann, D. and Gluckman, J. (1986) "HIV infection: Facts and hypotheses", Immunol. Today **7**:291-296.

24. Kowalsi, M., et al. (1987) "Functional regions of the envelope glycoprotein of human immundeficiency virus type 1", Science **237**:1351-1355.

25. Lifson, J., et al. (1986) "AIDS retrovirus induced cytopathology: Giant cell formation and involvement of CD4 antigen", Science **232**:1123-1127.

26. Maddon, P., et al. (1986) "The T4 gene encodes the AIDS virus receptor and is expressed in the immune system and the brain", Cell **47**:333-348.

27. Marsh, M. and Dalgleish, A. (1987) "How do human immunodeficiency viruses enter cells?", Immunol. Today **8**:369-371.

28. Marsh, M. and Helenius, A. (1989) "Virus entry into animal cells", In: Advances in Virus Research, Vol. 36; edited by Maramosch, K., Murphy, F. and Shatkin, A., Academic Press, Inc. San Diego, CA.

29. McClure, M., Marsh, M. and Weiss, R. (1988) "Human immunodeficiency virus infection of CD4-bearing cells occurs by a pH-independent mechanism", EMBO J. **7**:513-518.

30. Miyake, Y. and Okada, Y. (1978) "Effect of cytochalasin D on fusion of cells by HVJ (Sendai virus)", Exptl. Cell Res. **116**:167-178.

31. Mizukkami, T., et al. (1985) "Binding region for human immunodeficiency virus (HIV) and epitopes for HIV-blocking monoclonal antibodies of the CD4 molecule defined by site-directed mutagenesis", Proc. Natl. Acad. Sci. (USA) **85**:9273-9277.

32. Moore, N., et al. (1978) "Interaction of Vesicular Stomatitis virus with lipid vesicles: Depletion of cholesterol and effect on membrane fluidity and infectivity", J. Virol. **27**:320-329.

33. Nishiyama, Y., et al. (1976) "Polykaryote formation induced by Vesicular Stomatitis virus in mouse L cells", J. Gen. Virol. **32**:85-96.

34. Okada, Y., et al. (1975) "Modification of cell membranes with viral envelopes during fusion of cells with HVJ (Sendai virus) I. Interaction between cell membranes and virus in early stages", Exptl. Cell Res. **93**:368-378.

35. Pauza, C.D. (1988) "HIV persistence in monocytes leads to pathogenesis and AIDS", Cellular Immunology **112**:414-424.

36. Pauza, C.D., Bleil, J. and Lennox, E. (1984) "The control of transferrin receptor synthesis in mitogen-stimulated human lymphocytes", Exptl. Cell Res. **154**:510-520.

37. Pauza, C.D. and Galindo, J. (1989) "Persistent human immunodeficiency virus infection of monoblastoid cells leads to accumulation of self-integrated viral DNA and to production of defective virions", J. Virol., *in press*.

38. Pauza, C.D. and Price, T. (1988) "Human immunodeficiency virus infection of T cells and monocytes proceeds via receptor-mediated endocytosis", J. Cell Biol. **107**:959-968.

39. Popovic, M., et al. (1984) "Detection, isolation and continuous production of cytopathic retroviruses (HTLV-III) from patients with AIDS and pre-AIDS", Science **224**:497-500.

40. Robinson, W., Montefiori, D. and Mitchell, W. (1988) "Antibody-dependent enhancement of human immunodeficiency virus type 1 infection", Lancet **i**:790-794.

41. Sarin, P., et al. (1985) "Effects of a novel compound (AL721) on HTLV-III infectivity in vitro", New Engl. J. Med. **313**:1289-1290.

42. Stein, B., et al. (1987) "pH-independent HIV entry into CD4-positive T cells via virus fusion to the plasma membrane", Cell **49**:659-668.

43. Sundstrom, C. and Nilsson, K. (1976) "Establishment and characterization of a human histiocytic lymphoma cell line (U937)", Int. J. Cancer **17**:565-577.

44. Weyand, C., Gorozny, J. and Fathman, G. (1987) "Modulation of CD4 by antigenic activation", J. Immunol. **138**:1351-1354.

45. White, J. and Helenius, A. (1980) "pH-dependent fusion between the Semliki Forest virus membrane and liposomes", Proc. Natl. Acad. Sci. (USA) **77**:3273-3277.

46. White, J., Matlin, K. and Helenius, A. (1981) "Cell fusion by Semliki Forest, Influenza and Vesicular Stomatitis viruses", J. Cell Biol. **89**:674-679.

47. Willey, R., et al. (1988) "In vitro mutagenesis identifies a region within the envelope gene of the human immunodeficiency virus that is critical for infectivity", J. Virol. **62**:139-147.

## DISCUSSION

Mills (UCSF): What happens if you put your U937 cells in a recombinant soluble CD4 bath. Do you still see the phagosomes with what appear to be virus particles in them? If it was phagocytosis, you would anticipate that you would.

Pauza: We have done binding experiments, but we have not done electron microscopy since we obtained soluble CD4.

Fuller (University of Michigan): We work on herpes simplex virus entry, and I'm fascinated by the discussion of HIV entry either by receptor-mediated endocytosis or direct fusion. I think that it could probably happen both ways, and it depends on the cell that you're looking at. I have a couple of questions. First of all, in your first micrograph, there is a virion that you say looks like it could be fusing? At least you say it doesn't look like it's fusing, but to me it looks like there's continuity being formed there. Have you ever looked at viruses like this that have for some reason been treated with neutralizing antibodies that allow attachment, but do not allow penetration? In our system, we find lots of viruses stuck to the cell surface, but arrested at the fusion step. Have you performed such experiments with HIV and do you have such antibodies?

Pauza: No. I believe there is such an antibody, reported in a paper by Willey, et al. (1988; J. Virol. 62:139) and by Linsley et al. (1988; J. Virol. 62:3695) that block a post-adsorption step.

Fuller: This might be an interesting way to see where the virus has stopped, whether it will be there at the cell surface or in the endosome. It is known that (and I think this is from Ari Helenius' work), if you incubate cells at 4° and increase the temperature, you can increase phagocytosis or endocytosis. My question is do you have any markers for the vesicles to check that they really are endosomes, versus lysosomes or some other organelle? That is, do you know that the vesicles, where you have the viruses, are really endosomes, and have you done any work to look at that?

Pauza: We have not assessed the biochemical markers. The main thing that we rely on is the structural difference between the two forms of vesicles. One type always includes multiple particles, and the other one always incorporates single particles. The proposed endosomes often do not look like coated vesicles, though I do not know if that is surprising or not.

Fuller: So you think these are clathrin-coated pre-endosomes that the viruses are getting into?

Pauza: That is pushing the data too far.

Fuller: My last question: You showed the beginning of fusion out of the endosome. Do you ever see the next step, where you have the nucleocapsid right after the fusion event, so you can see that the capsid is there and there is the viral envelope, deposited either on the endosomal membrane, or on the cell surface? I don't know if this works for HIV, but you can see it with herpes simplex virus.

Pauza: I do not know of any evidence in the literature that pertains to that problem of whether you can see a retrovirus nucleocapsid free in the cytoplasm. My impression from the Semliki Forest virus studies, is that the nucleocapsid never leaves the vesicle and the replication complex is integral to the membrane. I might mention something that we tried that did not work. In early studies, we tried the lysosomotropic agents extensively. In kinetic studies, these agents slowed the whole process of HIV endocytosis, though not enough to establish that entry was pH-dependent. The effect looked more like a toxic effect of these compounds on cells. The second thing we tried was a variety of protease inhibitors, based on the work of Anderson and Nexø in Denmark. To date, we found a few protease inhibitors that seem to activate the infection a bit, but none that inhibit the infection.

Levy (UCSF): There are two premises that you make that we can draw on from the mouse system, to suggest maybe some other conclusions. The first is, when you get fusion or endocytosis and the core is released into the cytoplasm, the RNA, (your pictures you show it as free RNA) is always within the core and all reverse transcription takes place in the core. We know that because you can precipitate cores with antibody and you find the intact RNA, and the reverse transcription products. So the nature of the core and how it stays in the cytoplasm is not known; whether it has to be in vesicles or whether it can just exist as a free core within it is not known. One wonders whether all your studies of $^{32}P$ resistance would indicate that you have just got cores that have not been digested, because they are the survivors that will lead to infection. And that gets to the second question. As I recall in your studies, you use a ton of virus to get infection, to get your electron microscopy pictures, for instance. Since you have 300 defective to infectious particles, I am sure you are aware that the criticism would be that you are seeing the majority of particles which are defectives that may be handled quite differently from the infectious particles. Because, if you get a particle attached to CD4 and it is infectious, it may fuse very well, because of the dynamic situation. When you get defective particles, they sit there so long that endocytosis takes place. So, I think it is tough to get at, because we know all retroviruses exist with about 100-fold more non-infectious particles than infectious particles. It is a difficult situation, so we do go to biochemical analyses. So, while electron microscopy pictures are beautiful, they may merely be giving you the top of the picture.

Pauza: Absolutely. I think that the defective particle issue is going to be the single greatest stumbling block to a precise solution to this problem. I think we have taken the present approaches about as far as one can. We have considered other types of experiments. For example, recall the beautiful work from Ari Helenius' laboratory showing that anti-clathrin antibodies, introduced into cells via liposomal fusion, inhibit infection by Semliki Forest virus. This type of experiment would constitute a good approach to studying HIV entry. However, the SFV has a physical to infectious particle ratio of two. They utilize a plaque assay to obtain quantitative results and thus obtain a clean result concerning the degree of inhibition by anti-clathrin antibodies. I would like to know a better technical approach to solving the problem of virus entry for HIV. We suspect that endocytosis will also be the mechanism for virus entry when antibodies enhance the infection. It is important to understand the endocytic pathway of virus entry and to use this information to evaluate antiviral agents and neutralizing antibodies. Until some of these issues are resolved, I can't make solid and definitive arguments for the pathway of infectious HIV entry.

Larsen (UCSF): I had a question about what you called disruptive virions in endosomes. But first, I want to ask you about your Scatchard plot. You had on the x-axis counts from 50 to 1750. You don't say how many counts you start with, so I'd just like to have a sense of the percentage of bound virus versus what you've added.

Pauza: We don't use any preparations where less than 80% of the counts will bind to cells. So it has to be 80% or more of the added virus.

Larsen: Did you try any anti-CD4 to change one of those slopes?

Pauza: Well, we only did single-point pre-treatments (selecting one input virus concentration) with the anti-receptor antibody. The OKT4A antibody blocks binding to the CEM cells. These assays were done only in a high affinity range of virus binding. We didn't bother to examine the low affinity range of binding.

Larsen: Now, on the disrupted virions: You were answering the question by saying you didn't think it might be possible to see the nucleocapsid once it's inside the cytoplasm and outside of the endosome, once it's passed through the endosome, presumably it's still in some form.

Pauza: I just don't know...

Larsen: Okay. Well, you call it disrupted inside the endosome, which suggests that it's a nuclear capsid already. It seems the only thing that could be disrupted is the membrane anyway.

Pauza: Yes, to me, that's what it looks like. I call them structurally disorganized particles for want of a better description. That was an attempt to be careful and prudent about interpreting the micrographs. Now, whether or not that represents a chemical/structural change, that's what I don't know.

Fuller: We also see a similar sort of structural change in Herpes virus that's entering the cells versus the progeny virus that's been synthesized. That is, progeny virus has a nice round conformation that looks real great. In contrast, virus that's infectious is going into the cell and we can clearly show it's infectious virus, because we do have neutralizing antibodies that correlate biochemical activity with what we see by electron microscopy, and the virus that's going in has a very different orientation of its envelope and somehow seems structurally changed. We think this may be due to the fact that glycoproteins have to change conformationally as the virus goes in. This happens at the plasma membrane, however. But, there is some evidence also with reovirus that there is a conformational change in the envelope during the infectious process, so what you may be seeing there may be a very real phenomenon. I don't know, but we do see it in other systems.

Banapour (UC Davis): I was just wondering if you could repeat what is the exact evidence that virus that's present in the phagosomes are not infectious?

Pauza: Yes. There's a little jump of logic in the middle of the argument.. One observes efficient binding and rapid internalization of labeled virions. Most of these virus particles are directed along the pathway to degradation. We cannot inhibit binding or internalization significantly by pre-treatment of the U937 cells with OKT4A antibody, although we can inhibit efficiently the infection. Therefore, we observe a dominant process if binding and internalization that is separable from the infectious route of entry. Since the majority of virions bind to U937 cells even in the presence of OKT4A antibody, we presume that the rest of the virions enter via the rapid internalization pathway and are directed toward the phagosomes. Thus phagocytosis occurs whether or not we block the receptor and it doesn't seem to be an infectious route of entry. Presently we are concerned about antibody enhancement of infection experiments. Based on our understanding of phagocytosis of HIV, we would predict that aggregated virus, or large antibody-virion comoplexes, would preclude seeing an enhancement effect in infection experiments. This is due to the fact that Fc-receptor positive cells would simply phagocytose the material because it is a large aggregate and the virus would be forced to enter the cell via a non-infectious route. In all these experiments, we are actually concerned about discrimination by the macrophage between infectious and non-infectious routes of virus uptake. This discrimination will determine whether the encounter between HIV and the macrophage results in establishment of infection. Remember that these cells are sites of infection and a primary component of antiviral defense mechanisms.

Galloway (UCSF): Two questions: In macrophages, proteolysis starts quite early after internalization, say five minutes. Have you been able to add protease inhibitors and stop, not infection, but the structural change that happens to virus, after internalizaton?

Pauza: That experiment has not been done.

Galloway: The second question. Have you been able to inhibit this lower affinity binding with mannose or mannans?

Pauza: No. When we were doing those experiments we just never thought to do it. When I initially looked at HIV binding curves, I was reminded of my time spent studying kinetics of allosteric enzymes. I would interpret the low enzyme activity as "bad material" in the preparation. Naturally, I assumed that low affinity binding was due to "bad virus" in the preparation. Currently, I think there are good reasons to go back and re-investigate virus binding in the low affinity range. Inhibiting binding with sugars, particularly mannose, would be appropriate experiments.

Debs (UCSF): I believe U937 cells are a primitive myelomonocytic line which lack several features of differentiated monocytes or macrophages. At least with the U937 clone that we use, we're unable to induce internalization by any means of particulate matter that is readily taken up by both peripheral blood monocytes and tissue macrophages. Have you compared your observations in U937 to isolated peripheral blood monocytes to confirm that that's the same pathway and mechanism?

Pauza: We've never done it by that comparison, although these cells readily phagocytose many things. They take up iron particles or latex beads for example. In our lab, we make a living on these cells, so we tend not to use the pejorative descriptive adjectives that you used. Nonetheless, we recognize their differences from normal cells, even though they are competent phagocytes.

Düzgünes (UCSF): Does the $^{32}P$ label that you use go into the RNA or the phospholipids, and wherever it goes, does the degradation of the $^{32}P$ alter any of your conclusions?

Pauza: No. We're usually doing those experiments within one or two days of preparing the labeled virus. The best estimates we have indicate that better than 90% of the label is in RNA; i.e., is ribonuclease-sensitive. If you lyse virions with NP-40, treat with ribonuclease and then recover macromolecular material by TCA precipitation, 90% of the counts appear RNase sensitive.

Düzgünes: Could you comment on why the viral particles that you see in the phagosomes won't infect; whether they are degraded too rapidly to have a chance to infect, and if they are CD4-dependent?

Pauza: Personally, I really don't understand the difference in those two pathways. It is important to make this distinction for each virus family. Many viruses are infectious via the phagosome. Why HIV is not, I don't yet know.

Fuller: In many of the viruses that enter by receptor-mediated endocytosis, the acidification is the critical step, and if you take those proteins and express them in cells, you can cause the cells to fuse by a low pH pulse, and I don't know if you would do this or if there are other people here that can. In cells that are expressing HIV proteins, does a low pH pulse cause any change in the fusogenic state? If the virus can get in by receptor-mediated endocytosis, which implies a low pH step, then you would expect to see this, I would think.

Pauza: Well, other people here are probably better able than me to answer that question. The data which have been confirmed many times now, are that cells don't require any help to fuse, they do it just fine in tissue culture, and that pre-treatment of virions with acid has no measureable effect on infectivity. Barry Stein will tell you the rest of the story. Acidification of endosomes down to around pH 5 or 5.5 is not necessary for HIV infection. One precedent for this result is the example of amantadine-resistant mutants of influenza virus. There you have flu variants that still seem to enter via the endosome and do not require acidification to low pH. That would be one of the only systems to look at for guidance in understanding this phenomenon.

Levy: I just want to comment that two very important features were brought up by this discussion and it's something we don't understand. First is endocytotic. We've done those experiments and yet the virus is infectious. So that, I'm sure, deals with the cell biology of phagosomes versus endosomes. I mean, there are enzyme differences that can easily be approached by looking at the sensitivity of a virus. The other approach is fusion. Attachment is not an energy-dependent phenomenon, so you can get attachment in the cold, but fusion takes energy. And if you do your experiments with killed virus, you don't know what you're going to do to that fusion phenomenon. Something else I meant to comment on is, when you label with $^{32}P$, you essentially kill the virus. We have done experiments like this and looked at infectivity I think what you are getting is virus taken up. I think it is resistant to RNAse because it's in its core until its broken down, but until you can work out a system in which you've got infectious virus, which we've discussed, the approach of looking at the mechanisms for infection versus just how the cell handles virions is not approachable.

# ROLE OF CD4 IN THE PENETRATION OF CELLS BY HIV

J.S. McDougal, P.J. Maddon, G. Orloff, P.R. Clapham, A.G. Dalgleish, S. Jamal, R.A. Weiss and R.A. Axel

Immunology Branch, Division of Host Factors, Center for Infectious Diseases, Centers for Disease Control, Public Health Service, U.S. Department of Health and Human Services, Atlanta, GA 30333

Department of Biochemistry and Molecular Biophysics and the Howard Hughes Medical Institute, College of Physicians and Surgeons, Columbia University, New York, NY 10032

Institute of Cancer Research, Chester Beatty Laboratories, London SW3 6JB, England

Clinical Research Centre, Northwick Park Hospital Harrow HA1 3UJ, England

Enveloped viruses penetrate host cells by two distinct mechanisms: by direct fusion between viral and cell membrane or by receptor-mediated endocytosis. Though not mutually exclusive, it is likely that viruses use one pathway or the other, that one predominates, or even that both occur but only one results in successful infection in the sense that productive infection ensues. Stein et al.[1] and McClure et al.[2] have shown convincingly that the low pH environment of the endosome is not necessary for infection by HIV. This does not mean that endocytosis does not occur; it may be that HIV is endocytosed but the low pH environment is not required for access of viral material to the cytoplasm. Electron microscopic studies have not been particularly helpful, and one can find examples supporting either mechanism[1,3]. Favoring the direct fusion mechanism is the observation that HIV-infected cells fuse with uninfected CD4 cells at neutral pH. The membrane fusion or syncytia phenomenon requires interaction between CD4 and viral envelope glycoprotein gp120 and its mechanism is likely analagous to infection by cell-free virus. However, the endocytosis mechanism has a certain natural appeal in that the CD4 molecule, the receptor for HIV, has been shown to be internalized in endosomes as a result of T cell activation by mitogen, antigen, or phorbol ester. This occurs as a consequence of protein kinase C (PKC) activation and is associated with phosphorylation of the cytoplasmic segment of CD4[4-7]. Bedinger et al.[8] and Maddon et al.[7] have made CD4 constructs lacking the cytoplasmic segment of CD4 that are incapable of PKC-induced internalization. Cells expressing these mutant CD4 molecules can be readily infected with HIV[7,8]. However, ligand binding may induce internalization of a receptor independent of mechanisms involving PKC-activated phosphorylation of the cytoplasmic segment of the receptor. The EGF receptor is a case in point. Mutations in the cytoplasmic segment of the EGF receptor rendering it unresponsive to PKC induced internalization do not affect ligand-induced internalization[9].

It should be possible to distinguish the two mechanisms of penetration by monitoring cell surface expression of CD4 during the penetration event. Herein, we present our attempts to track the CD4 molecule during HIV penetration. We also report the effect of inhibitors of various intracellular signal transduction pathways on penetration, attempts to enhance HIV infectivity by inducing CD4 internalization, and genetic experiments with mutant CD4 molecules. We fail to find any evidence that CD4 internalizes with virus, nor does HIV penetration require signal transduction involving the cytoplasmic segment of CD4.

As a preface, two observations are fundamental to the design and manipulations of the experiments that follow. The first observation is that treatment of cells with appropriate doses of trypsin strips the external domains of CD4 from the cell. Trypsin-treated cells do not bind HIV, and the CD4 epitope recognized by OKT4A monoclonal antibody (mAb) is removed, although the cells still react with OKT4 mAb. OKT4 mAb reacts with a more membrane proximal epitope of CD4 that is not involved in HIV binding[10,11]. Cell-surface HIV cannot be detected on HIV-incubated and trypsin-treated cells. Furthermore, the supernate from an HIV-incubated and trypsin-treated cell preparation is noninfectious[11]. Thus, any infectious HIV derived from such a cell preparation will have to have penetrated at the time of trypsin treatment. The second observation relates to the fact that HIV binds to cells at 0°C but does not penetrate. When cells are incubated with serial dilutions of HIV for 2 hours, washed to remove unbound virus, plated in microcultures at 37°C, and virus production

Table 1. HIV Binding and Penetration at 0°C and 37°C

| HIV incubation 2 hours | Post-incubation treatment | Virus replication in microculture at 8 days $\log_{10}$ ID-50 titer |
|---|---|---|
| 0°C | -- | 3.31 |
| 0°C | trypsinize | <1.00 |
| 37°C | -- | 3.22 |
| 37°C | trypsinize | 3.53 |

measured, the registered ID-50 infectivity titer is similar to that of cells that were initially incubated with HIV at 37°C (Table 1). However, if the cells are trypsinized after binding at 0°C and before raising the temperature, infectivity is orders of magnitude lower. Clearly virus binds at 0°C because infectivity after washing is equivalent to the 37°C incubation, but, unlike the 37°C incubation, it remains accessible to trypsin. Thus, it should be possible to allow virus to accumulate on the cell surface at 0°C, and obtain a relatively synchronized penetration event when the temperature is raised to 37°C.

Experiments designed to track CD4 and HIV during the penetration event were performed as follows. Cells were preincubated at 0°C with HIV to allow accumulation of bound virus. Two types of HIV preparations were used: an HIV-containing culture supernate with high infectivity titer and a concentrated HIV preparation[10,11]. Both were used in saturating amounts in that they blocked detection of the OKT4A epitope of CD4 on the cells. Experiments were performed with phytohemagglutin (PHA)-stimulated peripheral blood lymphocytes (PBL), the CD4-positive cell line CEM, and HeLa cells transfected with CD4[12]. After the 0°C preincubation, the temperature was raised to 37°C, and CD4 and HIV density on the cell surface were monitored by cytofluorometry over time (Figure 1). We detected CD4 using the OKT4 mAb, a mAb that does not interfere with HIV binding. In comparison to control cells not incubated with HIV, we could not detect any perturbation in cell surface expression of CD4 (Figure 1). The phorbol ester TPA readily induced downmodulation of cell surface CD4.

To detect internalized CD4 or HIV in these same experiments, we trypsinized the cells and confirmed removal of cell surface CD4 and HIV by flow cytometry. Cells were then permeabilized and fixed for detection of cytoplasmic CD4 and

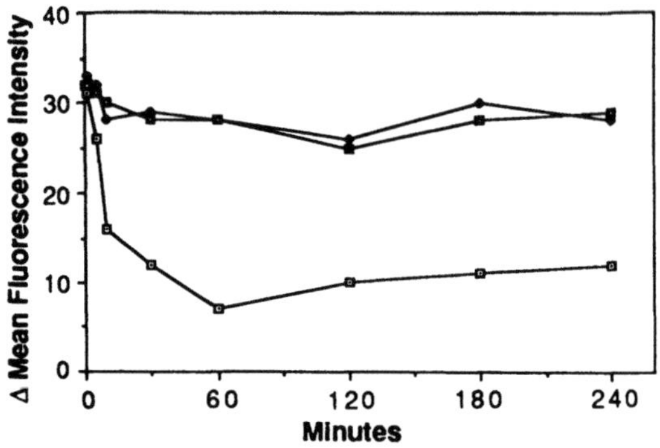


Fig. 1. Surface CD4 expression during HIV incubation. $CD4^+$ HeLa cell transformants were incubated with medium (◆——◆), with a saturating concentration of HIV (□——□), or with TPA (⊡——⊡) at 0°C for 1 hr. Cells were then incubated at 37°C. At various times, cells were removed to buffer at 0°C, and cell surface CD4 density determined by flow cytometry using OKT4 mAb. Reproduced from reference 7, with permission.

HIV. We used the OKT4E mAb for detecting internalized CD4 because reactivity of this mAb is not interfered with by HIV binding and the epitope with which it reacts is trypsin sensitive[11,13]. Intracellular HIV antigens were readily detected by this technique but intracellular CD4 was not. However, the quality of staining of intracellular CD4 in our positive control (TPA treated and trypsinized cells) was poor as well. In a different approach, we prelabelled cells with $^{125}$I-labelled OKT4 mAb and monitored trypsin-sensitive and trypsin-resistant radioactivity over time. $^{125}$I-labelled OKT4, when present on the cell surface, is trypsin sensitive. When CD4 is internalized by treatment with TPA, most of the label becomes trypsin resistant (Figure 2). The $^{125}$I-OKT4-CD4 complex on HIV-incubated cells remains trypsin sensitive when monitored at multiple time points over 4 hours. Thus, despite our best efforts to saturate the cells with HIV and synchronize the penetration event, we have not been able to detect any perturbation in cell surface CD4.

We have also tested whether the induction of CD4 internalization in HIV incubated cells would enhance infectivity. In this protocol, cells were incubated with serial dilutions of an HIV preparation. Four hours later TPA, was added and incubation was continued for another 4 hours. The cells were washed, plated in microculture, and infectivity titers determined. There was no enhancement of infectivity (Table 2). If anything, infectivity was lower in the TPA-treated cells.

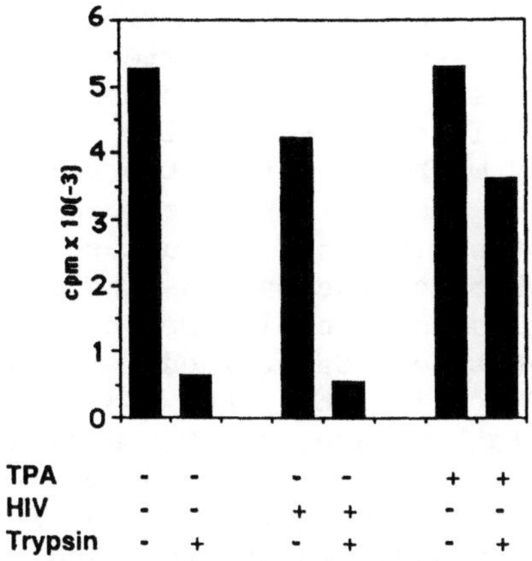


Fig. 2. Effect of CD4 binding on CD4 internalization. CD4+ cells were incubated with $^{125}$I-OKT4 at 0°C for 1 hr. Cells were washed and incubated with medium, with HIV, or with TPA for 1 hr at 0°C as indicated. Cells were incubated at 37°C for 4 hr, washed, trypsinized as indicated, and cell-associated radioactivity determined. Reproduced from reference 7, with permission.

In the converse approach, we determined whether inhibition of CD4 endocytosis would affect HIV infectivity. The CD4 internalization that occurs as a consequence of T cell activation is dependent on PKC activation. There are isoquinoline derivatives that inhibit PKC activity[14]. We used the PKC inhibitor H7 in concentrations sufficient to prevent TPA-induced downmodulation of cell surface CD4 and equimolar amounts of a control isoquinoline derivative HA1004 that has a lesser effect on PKC but greater effects on other kinases[14]. In this protocol, H7 or HA1004 were added to cells before a 4 hour incubation with serial

Table 2. Effect of TPA (added after HIV) on HIV Infectivity Titer.

| Cells | $Log_{10}$ ID-50 titer TPA | control |
|---|---|---|
| PHA-stimulated PBL | 2.11 | 3.97 |
| Unstimulated PBL | 2.40 | 4.73 |
| CEM | 2.37 | 2.08 |

dilutions of HIV. The cells were then trypsin treated so that only penetrated virus (if any) will register in the subsequent ID-50 assay. The cells are incubated another 4 hours during which a second set of cultures receives the inhibitors. These cultures were incubated with HIV during the first 4 hours, and they serve as a toxicity control for subsequent replication in that they differ from the first set of cultures only in the sequence of HIV and inhibitor incubation--the so-called delayed addition controls. All sets of cells were then washed, plated in microculture, and ID-50 titers determined. Inhibition studies were conducted using PHA-stimulated PBL, CEM, and freshly obtained PBL. In the latter case, the PBL were activated with PHA and IL-2 after the incubations to enhance HIV replication in the microcultures. Titers were equivalent in H7-incubated, HA1004-incubated, delayed addition control, and control cultures (Table 3). This assay registers ID-50 titers of a virus inoculum that has penetrated cells (rather than simply bound to them) at the end of the 4 hour incubation. The endpoint of the assay measures infectivity under conditions of limiting virus input and the 4 hour incubation is suboptimal for maximal infectivity. As such, it should be very sensitive to agents that inhibit or retard penetration. If there is no effect on the assay, we conclude there is no effect on penetration. On the other hand, if there is an effect, additional controls or experiments are required to determine whether the effect is on prepenetration events (binding) or postpenetration events (reverse transcription, integration, replication).

It should be noted that others have reported that the PKC inhibitor H7 inhibits HIV replication when continuously present in culture[15]. We have confirmed this observation. These results would indicate that PKC activity is required at some phase of the viral replication cycle, but is not required during the binding and penetration phase.

Table 3. Effect of PKC Inhibitors on HIV Infectivity for PHA-stimulated Peripheral Blood Lymphocytes.

| HIV incubation 1st 4 hr | 2nd 4 hr | HIV titer $Log_{10}$ ID-50 |
|---|---|---|
| H7 + HIV | -- | 4.06 |
| HIV | H7 | 3.98 |
| HA1004 + HIV | -- | 4.07 |
| HIV | HA1004 | 4.07 |
| HIV | -- | 3.87 |

Genetic manipulations of the CD4 molecule have provided even more convincing evidence that CD4 internalization (via a pathway requiring PKC-dependent phosphorylation of the cytoplasmic segment of CD4) is not required for HIV penetration. Mutations were introduced into CD4 that rendered the molecule incapable of internalization in response to PKC activation[7]. For one of the mutants, a stop-codon was introduced in the nucleotide sequence corresponding to amino acid position 405, just inside the cytoplasmic domain, resulting in a cytoplasmic deletion mutant. The other mutant had a substitution of alanine for serine at position 410. The serine-410 is surrounded by a putative consensus sequence for PKC phosphorylation. Both mutants were transfected and expressed in HeLa cells. In comparison to HeLa cells containing wild-type CD4, CD4 expression is downmodulated by PKC activation in the wild-type but not in mutants. Wild-type CD4 is phosphorylated as a consequence of PKC activation; the cytoplasmic deletion mutant is not; and phosphorylation of the serine-410 mutant is reduced about 80%. All three cell lines bind virus and have similar degrees of CD4 expression (clones were selected for similar degrees of CD4 expression). These mutant CD4 cell lines, incapable of PKC-induced internalization, were then tested for HIV infectivity by a variety of assays including quantitative infectivity assays. HIV infectivity was similar in all three cell lines (Table 4).

A function for CD4 in other related signal transduction pathways has been postulated as well. CD4 may be involved in the regulation of $Ca^{++}$-dependent fluxes that occur during activation in that mAb to CD4 perturb an expected $Ca^{++}$ influx, as does HIV, in some systems[16-18]. With the same infectivity protocol used to assess the effect of PKC inhibitors on HIV penetration, we have determined whether $Ca^{++}$ chelation or cyclosporin A inhibit HIV penetration. EGTA, in concentrations sufficient to chelate extracellular $Ca^{++}$, did not affect infectivity in this assay. Cyclosporin A does not prevent $Ca^{++}$ influx; rather, it inhibits certain activation events that are dependent on $Ca^{++}$ mobilization[19,20]. Cyclosporin A, in doses sufficient to inhibit mitogen-induced proliferative responses, had no effect in the HIV penetration-infectivity assay.

In summary, we can find no evidence that CD4 is internalized with HIV or that the capacity for CD4 internalization (at least by a PKC-dependent pathway or a pathway requiring phosphorylation or an intact cytoplasmic segment of CD4) is required for penetration. Nor can we find any requirement for influx of extracellular $Ca^{++}$ or cyclosporin A-sensitive activation events dependent on $Ca^{++}$ mobilization. It seems unlikely but remains possible that HIV ligand binding to CD4 induces CD4 internalization by a pathway not involving

Table 4. Susceptibility of Mutant $CD4^+$ HeLa Transformants to HIV Infection.

| Human Cells | Maximum reverse transcriptase | Cytoplasmic virus | Supernatant viral antigen | Supernatant subculture | Syncytium induction | VSV(HIV) pseudotype infection | HIV binding |
|---|---|---|---|---|---|---|---|
| CEM ($CD4^+$) | 1,536,780 | + | + | + | + | + | + |
| HeLa | 4,350 | – | – | – | – | – | – |
| HeLa-$CD4^+$ | 147,291 | + | + | + | + | + | + |
| HeLa-$CDcyt^-$ | 188,842 | + | + | + | + | + | + |
| HeLa-CDser410 | 168,427 | + | + | + | + | + | + |

Reproduced from reference 7, with permission.

these elements and that CD4 is rapidly returned to the membrane, an event we may not have been able to detect despite our best efforts to saturate cells with HIV and synchronize the penetration event.

The sole function of CD4 in the penetration of HIV may be to focus HIV on the cell surface. This is not a trivial function. The high affinity interaction of HIV with CD4 may serve to overcome the repulsive forces that exist between viral and cell membranes. Once the virus is focused in close proximity to the membrane, fusion events determined by viral or cell membrane components may then occur.

## REFERENCES


1. B. S. Stein, S. D. Gowda, J. D. Lifson, R. C. Penhallow, K. G. Beusch, and E. G. Engleman, pH-independent HIV entry into CD4-positive T cells via virus envelope fusion to the plasma membrane, Cell 49:659 (1987).
2. M. O. McClure, M. Marsh, and R. A. Weiss, Human immunodeficiency virus infection of CD4-bearing cells occurs by a pH-independent mechanism, EMBO J. 7:513 (1988).
3. C. D. Pavza and T. M. Price, Human immunodeficiency virus infection of T cells and monocytes proceeds via receptor-mediated endocytosis, J. Cell Biol. 107:959 (1988).
4. R. B. Acres, P. J. Conlon, D. Y. Mochizuki, and B. Gullis, Rapid phosphorylation and modulation of the T4 antigen on cloned helper T cells induced by phorbol myristate acetate or antigen, J. Biol. Chem. 261:16210 (1986).
5. J. A. Hoxie, D. M. Matthews, K. J. Callahan, D. L. Cassel, and R. A. Cooper, Transient modulation and internalization of T4 antigen induced by phorbol esters, J. Immunol. 137:1194 (1986).
6. J. A. Hoxie, J. L. Rackowski, B. S. Haggarty, and G. N. Gaulton, T4 endocytosis and phosphorylation induced by phorbol esters but not by mitogen or HIV infection, J. Immunol. 140:786 (1988).
7. P. J. Maddon, J. S. McDougal, P. R. Clapham, A. G. Dalgleish, S. Jamal, R. A. Weiss, and R. Axel, HIV infection does not require endocytosis of its receptor, CD4, Cell 54:865 (1988).
8. P. Bedinger, A. Moriarty, R. C. von Borstell, N. J. Donovan, K. S. Steimer, and D. R. Littman, Internalization of the human immunodeficiency virus does not require the cytoplasmic domain of CD4, Nature 334:162 (1988).

9. C. R. Lin, W. S. Chen, C. S. Lazar, C. D. Carpenter, G. N. Gill, R. M. Evans, and M. G. Rosenfeld, Protein kinase C phosphorylation at Thr 654 of the unoccupied EGF receptor and EGF binding regulate functional receptor loss by independent mechanisms, Cell 44:839 (1986).
10. J. S. McDougal, M. S. Kennedy, J. M. Sligh, S. P. Cort, A. Mawle, and J. K. A. Nicholson, Binding of HTLV-III/LAV to T4+ T cells by a complex of the 110K viral protein and the T4 molecule, Science 231:382 (1986).
11. J. S. McDougal, J. K. A. Nicholson, G. D. Cross, S. P. Cort, M. S. Kennedy, and A. C. Mawle, Binding of the human retrovirus HTLV-III/LAV/ARV/HIV to the CD4 (T4) molecule: conformation dependence, epitope mapping, antibody inhibition, and potential for idiotypic mimicry, J. Immunol. 137:2937 (1986).
12. P. J. Maddon, A. G. Dalgleish, J. S. McDougal, P. R. Clapham, R. A. Weiss, and R. Axel, The T4 gene encodes the AIDS virus receptor and is expressed in the immune system and the brain, Cell 47:333 (1986).
13. P. E. Rao, M. A. Tulle, P. C. Kung, and G. Goldstein, Five epitopes of a differentiation antigen on human inducer T cells distinguished by monoclonal antibodies, Cell. Immunol. 80:310 (1983).
14. H. Hidaka, M. Inagaki, S. Kawamoto, and Y. Sasaki, Isoquinoline sulfonamides: novel and potent inhibitors of cyclic nucleotide dependent protein kinase and protein kinase C, Biochemistry 23:5036 (1984).
15. A. P. Fields, D. P. Bednarik, A. Hess, and W. S. May, Human immunodeficiency virus induces phosphorylation of its cell surface receptor, Nature 333:278 (1988).
16. P. M. Rosoff, S. J. Burakoff, and J. L. Greenstein, The role of the L3T4 molecule in mitogen and antigen-activated signal transduction, Cell 49:845 (1987).
17. G. P. Linette, R. J. Hartzman, J. A. Ledbetter, and C. H. June, HIV-1-infected T cells show a selective signaling defect after perturbation of CD3/antigen receptor, Science 241:573 (1988).
18. H. Kornfeld, W. W. Cruikshank, S. W. Pyle, J. S. Berman, and D. M. Center, Lymphocyte activation by HIV-1 envelope glycoprotein, Nature 335:445 (1988).
19. P.M. Colombani, A. Robb, and A. D. Hess, Cyclosporin A binding to calmodulin: a possible site of action on T lymphocytes, Science 228:337 (1985).
20. P. M. Rosoff and G. Terres, Cyclosporine A inhibits $Ca^{++}$-dependent stimulation of the $Na^+/H^+$ antiport in human T cells, J. Cell. Biol. 103:457 (1986).

## DISCUSSION

Question: I have a question about the experiments with HeLa cells and the deletion mutants. You may have said this, but was the phosphorylation of protein kinase C looked at in the HeLa cells or was it looked at in another cell line?

McDougal: It was looked at in HeLa.

Question: Did you see internalization of the CD4 in response to phorbol esters?

McDougal: No, not with either the single point mutation or with the cytoplasmic mutant.

Question: The wild type?

McDougal: With the wild type, yes. The wild type does internalize. Expressed in HeLa, it does internalize with phorbol ester. And wild type also internalizes when expressed in mouse cells.

Pauza (Salk Institute): Steve, the virus still got in sometimes, right? So is this a hand-off mechanism? I mean is there some other protein carrying it in? You would be able to make everybody happy in one shot if it was.

McDougal: Well I think the possibility that there is a second phenomenon required for infection of human cells is very real. Whether it's an attachment phenomenon is not clear, but clearly there's a functional phenomenon. The best evidence for that I think, is really what Jay Levy was referring to in mouse cells. Mouse cells with proviral HIV DNA will make infectious progeny. Mouse cells expressing human CD4 cannot be infected. Therefore, the problem has to be either in penetration or in events that lead up to integration. It's likely that the problem is in penetration, based on the pseudotype experiments, where a pseudotype between HIV envelope and VSV would not infect these mouse cells, although VSV on its own will, make it likely that the problem is in penetration rather than some problem with reverse transcription in mouse cells.

Pauza: I'm sure you're probably aware of this, but the old literature on CD4 down-modulation was confused, saying that primary antibodies could never modulate CD4. A primary antibody plus a secondary antibody, crosslinking the whole mess into some giant aggregate was required to down-modulate. So, just to be clear, you've seen no rate of CD4 cycling in these cell lines?

McDougal: No, not with the $^{125}I$ experiments. We've not seen it in HIV-incubated cells, or in $^{125}I$-labeled CD4 cells, any diminution over the time span in the amount that's there. Now, it's very possible. It is cycled rapidly and returned to the cell surface. We're looking at density with a fluorescence assay.

Pauza: If you are using fluorescein-conjugated antibodies, and they go to endosomes, this should give a spectral shift. The emission spectrum of fluorescein is pH-dependent. If it was cycling, you should be able to see it.

McDougal: Yes, although that's really not quite the way we did it. We have actually tried to use fluorescein as well as the $^{125}$I-labeled antibody, but those were stained for. In the fluorescein experiments, they were incubated with HIV or not. And then, after whatever time point, we stopped it. Then they were stained with FITC anti-T4. But we have tried using FITC anti-T4 coated cells, to see if we can get the FITC label inside the cells by trypsinizing. We've had a lot of problems with that. Direct FITC-conjugated OKT4 is very trypsin-resistant. Probably because FITC labels lysines.

Stein (Stanford): I just want to go back to real basic cell biology outside of the realm of HIV. I spent quite a few years working on the transferrin receptor, and showed that there were two reclycing pathways for the receptors of the Golgi apparatus, and without even using HIV infection, you can't make a tacit assumption that both receptor pathways exist for all receptors. But in this particular case, you could test the hypothesis of normal CD4+ T-cells, treat them with phorbol ester (TPA, or tetradecanoyl phorbol acetate), and then put monensin in and see how long it takes for the receptor to come back to the surface. Because it may not recycle. It may be degraded like some of the receptors that are signal transducers. So you know to look at the problem more pragmatically, before attacking HIV, to see whether it is recycled. Now my experience with looking at CD4 is that monensin has absolutely no effect at all after five to seven hours on the number of CD4 receptors on T-cell lines, as opposed to some classic receptors like the transferrin receptor, EGF, and so forth, where if you treat with monensin or some other lysosomotropic agents, you'll see a reduction in cell surface receptors secondary to inhibition of recycling. So, maybe it's possible just to look at CD4+ cells to see if recycling occurs at all.

McDougal: That's a very good suggestion, but in your experience, it doesn't recycle?

Stein: I didn't see any evidence for recycling, but it doesn't mean there is a monensin-resistant pathway. Fifty percent of the transferrin receptors will be blocked up in the Golgi apparatus. In the trans-Golgi on monensin treatment, the other 50% continue to recycle independent of the drug, so they have two recycling pathways.

Sodroski (Harvard): I was wondering if anybody has looked at other species expressing human CD4 besides human and mouse. Have simian cells been looked at, or other cell types?

McDougal: I don't know. As we discussed it, even the somatic cell hybrid experiments, between human and mouse cells, have not really been all that helpful in terms of deciding whether the problem with mouse cells is a dominant feature of mouse or the requirement for an accessory function of human cells. It's my understanding that somatic cell hybrids between hamster and human cells are poorly infectable as well.

Sodroski: So hamster cells expressing human CD4 aren't infectable.

McDougal: Yes, I think that's true.

Sodroski: Has anybody looked at simian cells at all, do you know?

McDougal: With human CD4? Not that I know of.

Cheng-Mayer (UCSF): Were the cells used mouse fibroblasts?

McDougal: No, that's a Kappler and Merrick T-cell hybridoma line. Mouse T-cell hybridoma line.

Woodle (Liposome Technology, Inc.): In regard to this, Richard Axel, at a presentation at Stanford, indicated that he'd tried nine different species unsuccessfully. I don't recall whether he used a simian cell line or not.

Sodroski: Is it universally true that human cell lines with human CD4 all work? I guess that's the corrollary to the question.

McDougal: Do you mean whether they all work in the sense that they can be infected? Productively infected? Well, all that we've looked at, which is about four, can be.

Landau (UCSF): I believe that a lot of different types of simian cells have been looked at for HIV infection, and even just expressing the endogenous simian CD4 proteins, they are infectable. However, we have tried CHO cells that express human CD4 and they're not infectable.

McDougal: I would like to make a comment about the activation versus non-activation. Except for the syncytia phenomena, I would interpret our results, if I could reconcile our results with the unintegrated DNA results, I would suggest that in our assay in non-activated cells, we are getting penetration during the first four hours. And then we trypsinize, so we're dealing with cells that have been penetrated by the virus. Then with the activation, we activate the reverse transcription of the virus, and therefore, in our assay which has a later readout of productive viral infection, we get infection. Whereas in your assay for however many hours, you do not get the readout. Now that all makes sense, except it does not explain the syncytia phenomena at all, and that's a major problem. I do confess I guess that non-activated cells are less efficient, either under conditions of limiting virus input which is what I showed mostly the ID50 assay, or under conditions where we use excess virus and then limiting-dilute the cells to determine the number, the frequency of infected cells. Non-activated cells are less efficient. But, I would argue it's a quantitative rather than a qualitative difference.

Pauza: Why didn't you get increased virus production in TPA-treated T-cells anyway via direct activation of the LTR?

McDougal: I don't know. The first experiment is not a great experiment. The CD4 internalization is not the only thing going on with the phorbol ester-induced down-modulation. In fact, even if we'd gotten enhanced infection, one could argue that what we really did was super-activate the cells rather than super-internalize the virus.

# CD4 ENVELOPE INTERACTIONS OF HIV-1 AND RELATED HUMAN RETROVIRUSES

James A. Hoxie

Hematology-Oncology Section, Hospital of the University of Pennsylvania
3400 Spruce Street
Philadelphia, PA. 19104

The externalized envelope molecules of all human and simian immunodeficiency viruses thus far described have been shown to bind to the CD4 molecule expressed on the surface of some T lymphocytes and monocytes. The interaction of the envelope with CD4 is important both early in infection where it mediates the initial adsorption of the virion to the cell surface, and later during viral replication where it may contribute to a variety of cytopathic effects as well as the down modulation of the CD4 molecule from the cell surface. We have evaluated the extent to which interactions between CD4 and the viral envelope vary among different members of the HIV-1, HIV-2 and SIV family of viruses, and have attempted to determine if these differences correlate with biological heterogeneity. We describe that remarkable differences do exist in the extent to which CD4-envelope binding occur and in the degree of CD4 modulation during viral replication. These differences may contribute both to the cytopathic potential of particular viruses, as well as in their ability to establish superinfection interference of related viral isolates.

Several different strains of HIV-1 including IIIB, ARV-2 and WMJ1 were evaluated with respect to their host range on a variety of T cell lines, their ability to induce cytopathic effects including cell killing and fusion, and their effects on the expression of the CD4 molecule during infection. As shown in Table 1, cell free preparations of virus from all HIV-1 strains were able to infect the $CD4^+$ cell lines tested. Similarly, during the early phases of infection, both cell killing, as determined by trypan blue dye exclusion and fusion, as determined by the presence of multinucleated syncytia were observed, although the extent of these cytopathic effects varied somewhat depending on the particular viral strain and cell line utilized. We also evaluated CD4 expression on cells from these cultures which became chronically infected with the HIV-1 strains. Using immunofluorescence flow cytometry with monoclonal antibodies that have been shown either to compete (OKT4A) or not to compete (OKT4) with the HIV-1 envelope for binding to CD4[1,12] marked reductions in the levels of CD4 were seen on cell surface. This downmodulation of CD4 expression following HIV-1 infection has been described for several isolates and has been attributed to the complexing of CD4 with envelope precursor molecules within

*Mechanisms and Specificity of HIV Entry into Host Cells*
Edited by N. Düzgüneş, Plenum Press, New York, 1991

TABLE1. Infection of CD4+ cell lines by different strains of HIV and SIV

| | HIV-1 | | | HIV-2 | | SIV |
|---|---|---|---|---|---|---|
| CELL LINE | IIIB | ARV2 | WMJ1 | ROD | ST | MAC251 |
| HUT-78 | + | + | + | + | + | + |
| SUP-T1 | + | + | + | + | + | + |
| CEM | + | + | + | + | - | - |
| MOLT4 | + | + | + | + | - | - |
| JURKAT | + | + | + | + | ND | ND |
| CEMX174 | + | + | + | + | + | + |

Cell lines indicated were inoculated with cell free supernatant from HUT-78 cells productively infected with different strains of HIV-1, HIV-2 or SIV as previously described[18]. Cultures were monitored weekly for 28 days for the presence of reverse transcriptase (RT) in culture supernatants and for viral antigens by immunofluorescence microscopy (IFA) using sera from individuals infected with either HIV-1 or HIV-2. Serum with antibodies to HIV-2 was also used to detect SIVmac251 infection. Infection as evidenced by viral antigens on >10% of cells and RT levels >5 times background is indicated with a "+". In samples designated "-", no cells were positive by IFA and RT levels were not above background. ND, not done.

infected cells,[10,11] a reduction in the steady state level in CD4 mRNA,[10,13] translational defects in CD4 synthesis[13], and effects of the viral NEF gene product on cellular kinases that induce CD4 internalization[14].

In evaluating the relationship between binding of the HIV-1 envelope to CD4 and modulation of CD4 from the cell surface, we derived a biological variant of HIV-1IIIB that was unable to modulate CD4. Sup-T1 cells that were chronically infected with HIV-1IIIB were cloned by limiting dilution and evaluated for CD4 expression using both OKT4A and OKT4. As shown in Figure 1, the majority of clones (29 out of 30 assayed) exhibited a phenotype that resembled the chronically infected parental cells showing no reactivity to OKT4A and minimal to no reactivity with OKT4. However, one clone, designated A1.25, although unreactive with OKT4A, was strongly reactive with OKT4 (Figure 1).

The finding that this cell line expressed CD4 that could only be detected with a monoclonal antibody which did not compete with the HIV envelope for binding to CD4 indicated that the CD4 expressed by this cell line was most likely complexed with envelope but not downmodulated from the cell surface. Cells from the A1.25 line were 100% positive for viral antigens, as determined by immunofluorescence microscopy using serum from a patient with AIDS, and produced levels of reverse transcriptase activity in culture supernatants that were comparable to the uncloned parental line of HIV-$1_{IIIB}$-infected Sup-T1 cells (not shown). When supernatant from the A1.25 cell line was inoculated onto uninfected CEM and Sup-T1 cells, and chronically infected cultures established, high levels of surface CD4 were also detected on the productively infected cells as determined by OKT4 but not OKT4A. Thus, the lack of CD4 modulation in the A1.25 cell line appeared to reflect a property of the virus rather than the cell line used to

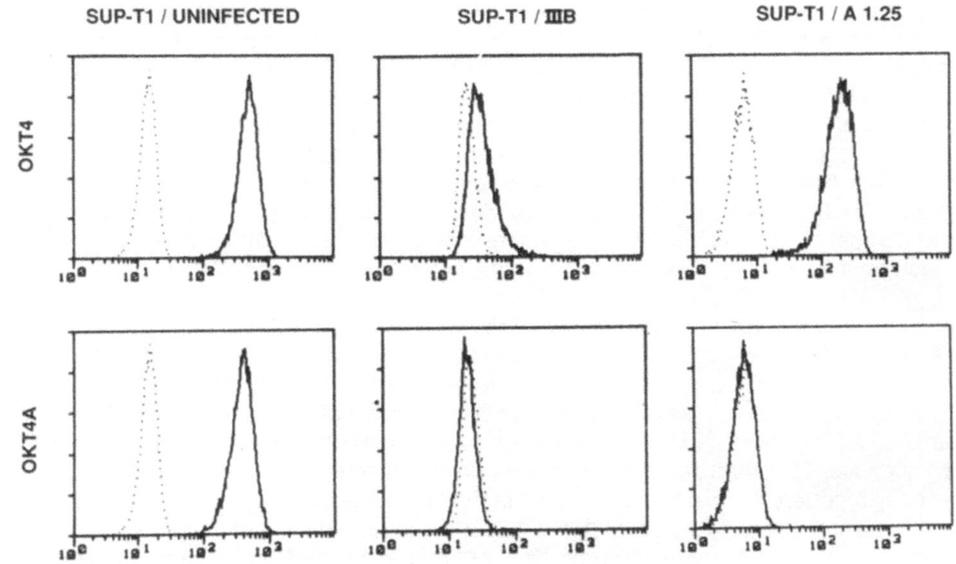


Figure 1. Modulation of CD4 induced by HIV-1IIIB and A1.25.
Sup-T1 cells chronically infected by the IIIB isolate of HIV-1 were cloned by limiting dilution and clones evaluated for CD4 expression using monoclonal antibodies OKT4 and OKT4A. Fluorescence histograms are shown for uninfected Sup-T1 cells, uncloned Sup-T1 cells infected by HIV-$1_{IIIB}$ (Sup-T1/IIIB), and a clone of infected cells that exhibit high levels of CD4 expression as determined by the OKT4 (A1.25).

produce the virus. In additional studies, we were able to precipitate complexes of CD4 gp120 and the gp160 envelope precursor molecule CD4 from $S^{35}$ cysteine- and $S^{35}$ methionine-labeled A1.25 cells, demonstrating that these envelope gene products were able to bind to CD4. Moreover, although A1.25 cells expressed high levels of surface CD4, these cells were completely resistant to fusion when cocultured with fusogenic strains of HIV-1 or HIV-$2_{ROD}$ (not shown), demonstrating that the viral envelope produced by the A1.25 cells was able to interfere with cytopathic effects mediated by the envelope of other HIV's.

These findings demonstrate that for HIV-1, binding of the viral envelope to CD4 and downmodulation of the CD4 molecule from the cell surface are not necessarily linked events. Our studies with the A1.25 biological variant of HIV-1 have demonstrated that an infectious virus can be obtained which binds to CD4 but fails to reduce the expression of CD4 from the cell surface following infection of other T cell lines. Therefore, while envelope binding to CD4 may play a role, it is not sufficient to downmodulate CD4. Other possible contributing factors include alterations in the interaction of the envelope or other viral gene products with CD4 or other cellular molecules that mediate expression of CD4 on the cell surface. Interestingly, Guy et al have demonstrated that the viral NEF protein which shows functional and structural similarity to cellular G proteins and the *ras* oncogene is able to produce modest reductions in CD4 expression when transfected into a T cell line[14]. Although it is possible that a reduction in NEF expression by the A1.25 virus may contribute to the defect this virus exhibits in inducing CD4 downmodulation, we have performed similar studies using a mutant of HIV-1, termed X10.1, which lacks the initiation codon for NEF [15] and produces no NEF protein[16]. Despite the absence of NEF, X10.1 is also able to induce a marked reduction in CD4 expression following infection of HUT-78 cells (not shown). Thus, it appears likely that a reduced level of NEF will not in itself explain the observed defect in the A1.25 virus. Molecular cloning and analysis of the A1.25 virus is in progress and will hopefully provide insight into the viral factors that are important in regulating CD4 expression on the cell surface following infection.

We also evaluated interactions of the viral envelope with CD4 and the effects of infection on CD4 expression for retroviruses related to HIV-1, including two strains of HIV-2 and the SIVmac251 strain of SIV. The ROD strain of HIV-2 isolated from a West African individual with AIDS[17], was found to be identical to strains of HIV-1 with respect to its host range (Table 1), its ability to induce killing and cell fusion during infection, and in the ability to modulate CD4 from the surface of chronically infected cell lines (not shown). As recently reported, SIVmac251 in contrast to the HIV-1's and HIV-2ROD exhibited a more restricted host range and was unable to infect or induce syncytia on cells such as CEM, Molt-4 and Jurkat even though the envelope of this virus was shown by immunoprecipitation analysis to bind to CD4 [18]. Interestingly, a hybrid cell line between CEM and an EBV transformed B cell line, termed CEMx174, was found to be highly susceptible to both infection and fusion by SIVmac251 indicating that molecules in addition to CD4 were likely required by this virus for infection and were either induced or provided to CEM by the B cell parent of this hybrid[18].

Although Sup-T1 cells were resistant to cell fusion when cocultured with SIVmac-producing cells, this cell line could be infected by cell free virus, but without any detectable syncytia formation, cell death, or other cytopathic effects[18]. Remarkably, in marked contrast to Sup-T1 cells infected by HIV-1 (or HUT-78 cells infected by SIVmac) the SIVmac-infected Sup-T1 cells remained strongly positive for CD4 as determined by both OKT4 and OKT4A indicating that viral-induced modulation of CD4 did not occur and that binding of the envelope to CD4 expressed on the surface of Sup-T1 cells was altered. Moreover, SIVmac-infected Sup-T1 cells, 100% infected by immunofluorescence microscopy, were highly susceptible to both cell fusion and infection when cultured with HIV-1 or HIV-$2_{ROD}$-infected cells or cell free virus, respectively, indicating that interference of HIV-1 or HIV-$2_{ROD}$ by SIVmac, at least on Sup-T1 cells, did not occur. Paradoxically, although the binding of OKT4A and four other anti-CD4 monoclonal antibodies was not inhibited on the SIVmac-infected Sup-T1 cells, we were able to precipitate complexes of envelope gene products with CD4 from metabolically labeled cells with OKT4 (but not OKT4A). Therefore, the SIVmac envelope is not absolutely defective in the ability to bind to CD4. It remains a possibility, however, that this envelope is impaired in its ability to bind to CD4 expressed on cells due to a reduced affinity for CD4 compared to HIV-1 or to effects of molecules associated with CD4 on the cell surface.

In addition to the differences noted above for the interaction of the SIVmac simian virus with human cell lines, we also observed interesting differences for a human isolate of HIV-2, termed HIV-$2_{ST}$. As recently described, HIV-$2_{ST}$ was isolated from a healthy Senegalese individual and was remarkable for its attenuated cytopathicity in vitro[19]. HIV-

$2_{ST}$ was noted to be poorly fusogenic and, in contrast to HIV-1 and HIV-$2_{ROD}$, produced little if any cell death during infection of cell lines or peripheral blood T cells. In subsequent studies, we have evaluated the ability of HIV-$2_{ST}$ to infect a number different T cell lines. As shown in Table 1, this virus was similar to SIVmac in its rather limited host range. Although HIV-$2_{ST}$ was able to infect lines such as HUT-78, Sup-T1, and the hybrid line CEMx174, it was unable to infect the CEM and MOLT4 cell lines. The HIV-$2_{ST}$ envelope bound to and coprecipitated with CD4, and CD4 expression on the surface of HIV-$2_{ST}$-infected Sup-T1 cells was reduced as demonstrated by reduced reactivity with both OKT4 and OKT4A. However, as shown in Figure 2, this downmodulation was not complete, and, in contrast to findings noted above with HIV-1 and HIV-$2_{ROD}$, detectable levels of OKT4A could still be demonstrated, despite the fact that these cells were 100% postive for viral antigens and produced high levels of reverse transcriptase activity in culture supernatants (>100,000 cpm).

In addition, HIV-2ST-infected Sup-T1 cells remained susceptible to fusion and superinfection by the IIIB and ARV-2 strains of HIV-1 as well as HIV-$2_{ROD}$ indicating that the HIV-$2_{ST}$ envelope was unable to completely interfere with other cytopathic strains of HIV-1 and HIV-2. In addition, the induction of cell fusion on HIV-$2_{ST}$-infected Sup-T1 cells by HIV-1 could be completely inhibited by the OKT4A and Leu-3a anti-CD4 antibodies indicating that CD4 was still utilized during infection. These findings for HIV-$2_{ST}$ are similar to those described by Evans and coworkers for another isolate of HIV-2, termed HIV-$2_{UC1}$, that failed to induce CD4 downmodulation and did not interfere with HIV-1 infection [20].

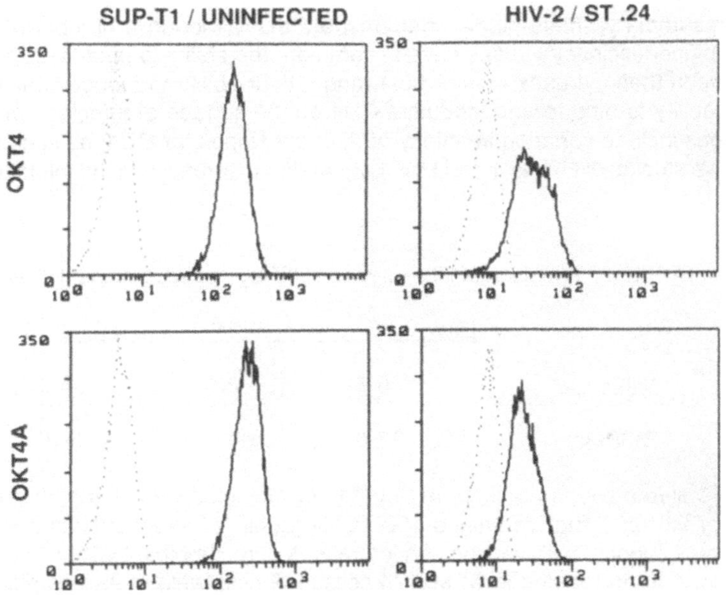


Figure 2. CD4 modulation induced by HIV-2ST.
Sup-T1 cells either uninfected or infected by HIV-2ST were evaluated for CD4 expression using anti-CD4 monoclonal antibodies OKT4 and OKT4A. The infected Sup-T1 cells were 100% positive for viral antigens as determined by immunofluorescense microscopy. In each panel, fluorescence histograms for the anti-CD4 antibodies (solid line) are superimposed over histograms for a nonreactive monoclonal antibody (dotted line).

The finding that the envelope of HIV-$2_{ST}$ could bind to CD4, but that Sup-T1 cells infected by HIV-$2_{ST}$ could be superinfected by HIV-1 and HIV-$2_{ROD}$ suggested that the HIV-$2_{ST}$ envelope could have a reduced affinity for the CD4 molecule compared to other more cytopathic strains of HIV-1 and HIV-2. In order to investigate this possibility, we determined the concentration of soluble CD4 required to inhibit infection by HIV-$2_{ST}$ and HIV-$1_{IIIB}$. Cell free supernatants containing 100 $TCID_{50}$ infectious units of HIV-$2_{ST}$ and HIV-$1_{IIIB}$ were precultured for 1 hr at 37° with varying concentrations of recombinant CD4 [20], kindly provided by Dr. Ray Sweet (Smith Kline Corp., King of Prussia, PA). The viral samples treated with CD4 were inoculated onto CEMx174 cells which, as noted above, are susceptible to infection by both HIV-$1_{IIIB}$ and HIV-$2_{ST}$. Cell cultures were passaged twice weekly and evaluated on days 7, 14 and 21 by immunofluorescence microscopy for the presence of viral antigens. As summarized in Table 2, in 3 experiments, $HIV_{IIIB}$ infection was completely inhibited by 0.5-4.0 µg/ml of soluble CD4. However, depending on the time at which infection was evaluated, significantly higher concentrations of recombinant CD4 were needed to inhibit HIV-$2_{ST}$ infection with concentrations ranging from 32 µg/ml on day 7 to 64 µg/ml on day 21. Similar findings have been reported by Clapham et al. for different strains of HIV-2 as well as SIV[22]. This 16 - 64 fold increase in the amount of recombinant CD4 that was required to inhibit HIV-$2_{ST}$ infection compared to HIV-$1_{IIIB}$ is consistent with the impression that there is reduced affinity of the HIV-$2_{ST}$ envelope for the soluble CD4 molecule. Studies are in progress to directly determine the affinity of the HIV-$2_{ST}$ envelope for CD4 using recombinantly expressed envelope of this virus produced by a molecular clone of HIV-$2_{ST}$ provided by Dr. Beatrice Hahn (University of Alabama, Birmingham, Ala)

In summary, these studies demonstrate that although a number of human and simian immunodeficiency viruses have in common, the ability to bind to CD4, differences exist between these viruses in their host range, their ability to induce viral interference, and their ability to bind to and modulate CD4 on the surface of infected cells. Possible explanations include variation in affinity of their envelopes for CD4, as suggested by our comparative studies of HIV-$2_{ST}$ and HIV-$1_{IIIB}$, and/or differences in the interaction of

TABLE 2. Inhibition of HIV-$1_{IIIB}$ and HIV-$2_{ST}$ by recombinant CD4.

| DAY- | 7 | 14 | 21 |
|---|---|---|---|
| HIV-$1_{IIIB}$ | 0.5 | 2.0 | 4.0 |
| HIV-$2_{ST}$ | 32.0 | 64.0 | 64.0 |

Cell free supernatants from Sup-T1 cells infected with either HIV-1IIIB or HIV-2ST that contained 100 $TCID_{50}$ units of infectious virus were preincubated with varying concentrations of recombinant CD4 [21] at 37° for 1 hr and added to CEMx174 cells ($10^5$ cells/well) in 96 well plates. Cells were washed after 24 hrs and immunofluorescence microscopy performed on the days indicated using sera from patients with either HIV-1 or HIV-2 infection. Numbers represent the concentration of CD4 (µg/ml) needed to completely inhibit viral infection at the indicated time points.

these viruses with other cellular molecules during binding, penetration and replication. Our observations with the A1.25 biological variant of HIV-1 as well as SIVmac and HIV-$2_{ST}$ have indicated that while the envelope-CD4 interaction may be necessary for viral binding, infection, and the subsequent downmodulation of the CD4 molecule from the cell surface, there are other viral and cellular factors which markedly affect these events. Finally, the finding that alterations in the CD4-envelope interaction of HIV and SIV viruses do not interfere with infection by some viral strains may have significant biological consequences in enabling recombination events to occur between genetically different viruses. Studies to evaluate the possibility that superinfection of HIV-$2_{ST}$ and SIVmac-infected cell lines by strains of HIV-1 or HIV-$2_{ROD}$ may lead to the generation of novel recombinant viruses with distinct biological properties are currently in progress. Such an event could have important implications in vivo as a mechanism for the evolution of diversity within this enlarging family of human and simian retroviruses.

Acknowledgments

This work was supported by Public Health Service grants UO1 AI2360-01 and PO1 AI25380-01 from the National Institutes of Allergy and Infectious Diseases, and a grant from Carter Wallace Inc. Dr. Hoxie is a Scholar of the Leukemia Society of America. We thank Drs. Beatrice Hahn and George Shaw (University of Alabama at Birmingham, Birmingham, Ala.) for providing the HIV-$2_{ST}$ virus, Dr. James Mullins (Department of Cancer Biology, Harvard School of Public Health, Boston, Mass.) for providing the SIVmac251 virus, and Dr. Phyllis Kanki (Department of Cancer Biology, Harvard School of Public Health, Boston, Mass.) for providing sera from individuals infected with HIV-2.

References

1. J. S. McDougal, M. S. Kennedy, J. N. Sligh, S. P. Cort, S. M. Kennedy, A. C.Mawle, Science 231:382 (1986).
2. M. Kowalski , J. Potz , L. Basiripour, T. Dorfman, W. C. Goh, E. Terwilliger, A. Dayton, C. Rosen, W. Haseltine, J. Sodroski, Science 237:1351 (1987).
3. L. A. Lasky, G. Nakamura, J. H. Smith, C. Fennie, L. Shimosaki, E. Patzer, P. Berman, T. Gregory, D. J. Capon, Cell 50:975 (1987).
4. D. H. Smith, R. A. Byrn, S. A. Marsters, T. Gregory, J. E. Groopman, D. J. Capon , Science 238:1704 (1987).
5. P. J. Maddon, A. G. Dalgleish, J. S. McDougal, P. R. Clapham, R. A. Weiss, R. Axel, Cell 47:333 (1986).
6. J. D. Lifson, G. R. Reyes, M. S. McGrath, B. S. Stein, E. G. Engleman, Science 232:1123 (1986).
7. J. Sodroski, W. C. Goh, C. Rosen, K. Campbell, W. A. Haseltine, Nature (London) 322:470 (1986).
8. A. DeRossi, G. Franchini, A. Aldovini, A. DelMistro, L. Chieco-Bianchi , R. C. Gallo, F. Wong-Staal, Proc. Natl. Acad. Sci. USA 4297 ( 1986).
9. R. Cheynier, M. Soulha , F. Loure, J. C. Vol, B. Reveil, R. C. Gallo, P. C. Sarin, D. Zagury , AIDS Res. and Human Retroviruses 4:43 (1988).
10. J. A. Hoxie, J. D. Alpers, J. L. Rackowski, K. Huebner, B. S. Haggarty, A. J. Cedarbaum, J. C. Reed, Science 234:1123 (1986).
11. M. Stevenson, C. Meier , A. M. Mann, N. Chapman, A. Wasiak, Cell 53:483 (1988).

12. Q. J. Sattentau, A. J. Dalgleish, R. A. Weiss, P. C. L. Beverley, Science 234:1120 (1986).
13. M. A. R. Yuille, M. Hugunin, P. John, L. Peer , L. V. Sacks, B. J. Poiesz, R. H. Tomar, and A. E. Silverstone, Jrl. Acquired Immune Deficiency Syndromes 1:131 (1988).
14. B. Guy, M. P. Kieny, Y. Riviere, C. Le Peuch, K. Dott, M. Girard, L. Montagnier, J-P. Lecocq, Nature 330: 266 (1988).
15. A. G. Fisher, L. Ratner , H, Mitsuya, L. M. Marselle, M. E. Harper, S. Broder, R. C. Gallo , F. Wong-Staal, Science 233: 655 (1986).
16. P. J. Durda, B. Leece, A. Jenoski, H. Rabin, A. Fisher, and F. Wong-Staal, AIDS Research and Human Retroviruses 4: 331 (1988).
17. M. Guyader, M. Emmerman, P. Sonigo, F. Clavel, L. Montagnier, and M. Alizon, Nature 326:662 (1987).
18. J. A. Hoxie, B. S. Haggarty, S. E. Bonser, J. L. Rackowski , H. Shan, and P. Kanki, J. Virol. 62: 2557 (1988).
19. L. I. Kong, S-W. Lee , J. C. Kappes, J. S. Parkin, D. Decker, J. A. Hoxie, B. H. Hahn, G. M. Shaw, Science 240:1525 (1988).
20. L. A. Evans, J. Moreau, K. Odehouri, H. Legg, A. Barboza, C. Cheng-Mayer, J. A. Levy, Science 240:1522 (1988).
21. K. C. Deen, J. S. McDougal, R. Inacker, G. Folena-Wasserman , J. Arthos, J. Rosenberg, P. J. Maddon, R. Axel , R. W. Sweet , Nature 82 (1988).
22. P. R. Clapham, J. N. Weber, D. Whitby, K. McIntosh, A. G. Dalgleish, P. J. Maddon, K. C. Deen , R. W. Sweet , and R. A. Weiss, Nature 337 (1988).

# FUSION OF HIV-1 AND $SIV_{mac}$ WITH LIPOSOMES AND MODULATION OF HIV-1 INFECTIVITY **

Nejat Düzgüneş* # §, Charles E. Larsen*, Krystyna Konopka*, Dennis R. Alford*, Lawrence J. T. Young^, Thomas P. McGraw^, Brian R. Davis@, Shlomo Nir* ‡ and Myra Jennings†

*Cancer Research Institute and #Department of Pharmaceutical Chemistry, University of California, San Francisco, CA 94143; §Department of Microbiology, University of the Pacific, School of Dentistry, San Francisco, CA 94115; ^California Primate Research Center and †Comparative Oncology Laboratory, University of California, Davis, CA 95616; @Medical Research Institute of San Francisco at Pacific Presbyterian Medical Center, San Francisco, CA 94115; ‡Department of Soil and Water Sciences, The Hebrew University of Jerusalem, Rehovot 76100, Israel

## INTRODUCTION

Human immunodeficiency virus (HIV) and simian immunodeficiency virus from macaques ($SIV_{mac}$) are the etiological agents of the acquired immunodeficiency syndrome (AIDS), and simian AIDS, respectively (Barré-Sinoussi et al., 1983; Gallo et al., 1984; Levy et al., 1984; Clavel et al., 1986; Letvin et al., 1985; DeVita et al., 1988). The genetic sequence of $SIV_{mac}$ is closely related to that of HIV-2 (Chakrabarti et al., 1987) and HIV-1 (Chakrabarti et al., 1987; Desrosiers, 1988), and infection of macaques by $SIV_{mac}$ represents the best animal model for human AIDS. The molecular and cellular mechanisms of entry of these viruses into their host cells are not well understood. The involvement of an endocytosis step before the fusion of the HIV-1 envelope with cellular membranes, and the requirement for acidic pH for fusion are topics of controversy (Maddon et al., 1986, 1988; Stein et al., 1987; Pauza and Price, 1988; McClure et al., 1988).

Although the CD4 molecule, found on circulating T helper cells, monocytes and macrophages, has been identified as the cellular receptor for HIV and $SIV_{mac}$ (Dalgleish et al., 1984; Klatzmann et al., 1984; McDougal et al., 1985; Kannagi et al., 1985), several cell types that do not express this protein can be infected by these viruses (Cheng-Mayer et al., 1987; Clapham et al., 1989; Tateno et al., 1989;

** Address for correspondence: Dr. N. Düzgünes, Department of Microbiology, University of the Pacific, School of Dentistry, San Francisco, CA 94115

Harouse et al., 1989; Weber et al., 1989; Konopka et al., 1991a). The expression of CD4 on the cell surface does not appear to be sufficient for HIV or $SIV_{mac}$ infection (Maddon et al., 1986; Koenig et al., 1989).

Viral envelope proteins are thought to participate in both the binding and fusion steps of the cell entry process (Düzgüneș, 1985; White, 1990; Ohnishi, 1988; Hoekstra & Kok, 1989; Stegmann et al., 1989a). The molecular basis for the fusion step is far less understood than for binding. The binding and fusion of immunodeficiency viruses are mediated by two envelope-associated glycoproteins. HIV-1 contains a trans-membrane glycoprotein, gp41, thought to mediate the fusion step, and a glycoprotein, gp120, attached non-covalently to gp41, and thought to mediate binding to CD4 (Lifson et al., 1986; Sodroski et al., 1986; McDougal et al., 1986a; Kowalski et al., 1987). HIV-2 and $SIV_{mac}$ contain two related envelope-associated glycoproteins, gp120 and transmembrane proteins that range in size from gp32 to gp40 (Chakrabarti et al., 1987). The N-terminal of gp41 (gp32/40) has some sequence similarity to the fusion proteins of other lipid enveloped viruses (Marsh & Dalgleish, 1987; Gallaher, 1987; Gonzalez-Scarano et al., 1987; Bosch et al., 1989), and is thought to interact with the target membrane bilayer and possibly act as a molecular bridge for the viral and cellular lipids. This interaction is also thought to destabilize the target membrane and increase the hydrophobicity of the membrane surface, thereby changing the water structure of the region of adhesion between the viral and target membranes and altering the energy of interaction between the two membranes (Ohki, 1988; Stegmann et al., 1989a; Düzgüneș and Shavnin, 1991). We have proposed that "this region of the viral envelope bilayer can no longer maintain its integrity when faced with a highly

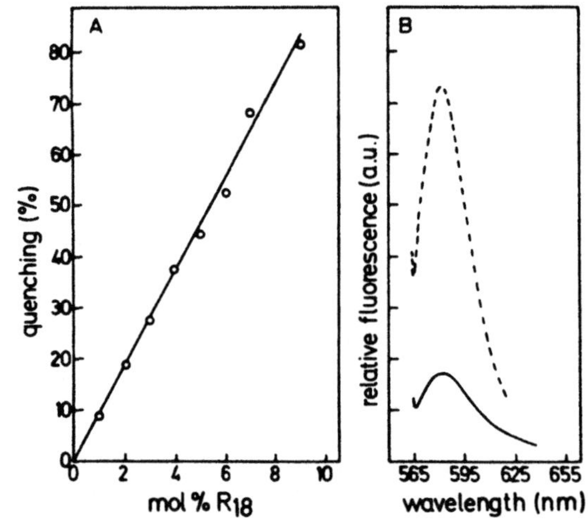


Figure 1. The efficiency of self-quenching of R-18 in dioleoyl-phosphatidylcholine vesicles (A), and the relief of self-quenching upon addition of detergent to Sendai virus containing R-18 in its envelope (B). The solid line in B is the the emission spectrum before the addition of detergent (excitation at 560 nm), and the dashed line is the spectrum obtained following detergent addition, with complete dilution of the probe. Reproduced with permission from Hoekstra et al., 1984.

hydrophobic surface with an altered water structure, and begins to adapt to the new physical boundary conditions, by altering its bilayer configuration into non-bilayer fluctuations. When two hydrophobic regions on apposed membranes coincide they coalesce to form a single membrane" (Düzgüneş and Shavnin, 1991).

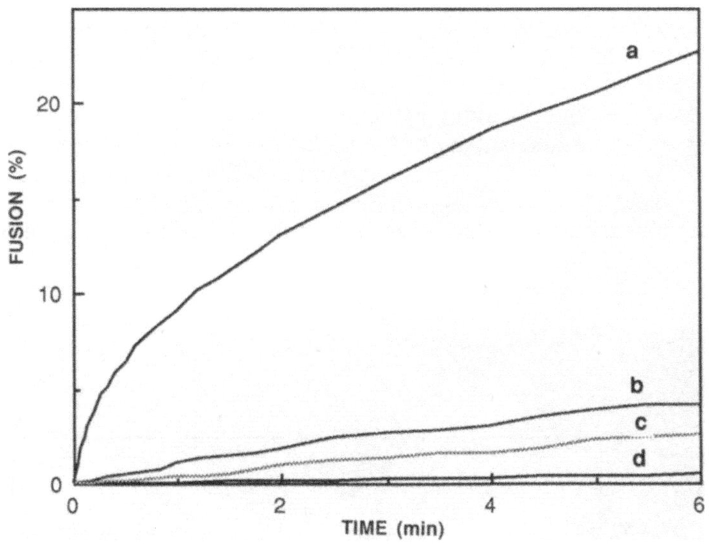


Figure 2. The time-course of fluorescence dequenching of R-18 incorporated into $SIV_{mac}$ (3 μg protein/ml) which was incubated at pH 7.5 at 37° C with large unilamellar liposomes (50 μM lipid) of varying composition. a) cardiolipin; b) cardiolipin/ dioleoylphosphatidylcholine (3:7); c) phosphatidylserine; d) dioleoylphosphatidylcholine. $SIV_{mac}$ was grown in HuT-78 cells incubated in RPMI 1640 medium and 5% fetal bovine serum in stationary cultures at 37° C (5% $CO_2$). The culture supernatant was clarified by low speed centrifugation and the virus was precipitated with saturated ammonium sulfate, pelleted at 6000 x g for 30 min, and resuspended in 10 mM Tris, 1 mM EDTA, 100 mM NaCl, pH 7.4 (TEN) buffer. The virus was purified by a 30 to 65% sucrose step gradient (275,000 x g for 30 min at 4° C) and a 5 to 65% continuous sucrose gradient (275,000 x g for 1 h 4° C) and stored at -70° C. Large unilamellar liposomes were prepared by reverse phase evaporation (Szoka et al., 1980; Düzgünes et al., 1981) in 150 mM NaCl, 10 mM *N*-[Tris (hydroxymethyl) methyl]-2-aminoethane sulfonic acid (TES), and 10 mM citrate, pH 7.5. Adapted from Larsen et al., 1990a, with permission.

## FUSION OF $SIV_{mac}$ WITH LIPOSOMES

To study the fusion of $SIV_{mac}$ and HIV-1 with target membranes, we have used the fluorescence dequenching assay developed by Hoekstra et al. (1984, 1985). The technique involves the insertion of the fluorescent probe, octadecylrhodamine (R-18), which enters spontaneously into viral membranes, and does not appear to exchange between membranes (Hoekstra et al., 1984; Loyter et al., 1988; Düzgünes et al., 1991a). The overlap in the excitation and emission spectra of R-18 allows for the phenomenon of self-quenching when the probe is inserted into a membrane at a concentration above a few mole percent of the lipid. When R-18 is incorporated into liposomes composed of dioleoylphosphatidylcholine (DOPC) at different mole ratios, the percent quenching is a linear function of the mol % R-18 (Figure 1; Hoekstra et al., 1984). The percent quenching is determined by comparing the fluorescence intensity of the liposomes with that obtained by lysing the liposomes with detergent to achieve infinite dilution of the probe. R-18 exhibits low fluorescence intensity when it is in the virus membrane. Upon fusion with a target membrane, the probe is diluted, and since self-quenching is dependent on the surface density of the probe, the fluorescence intensity of the fluorophore increases. The R-18 assay applied to influenza virus fusing with phosphatidylserine or cardiolipin liposomes as a target membrane, produced results similar to an alternative assay that measures the dilution of the phospholipid probes

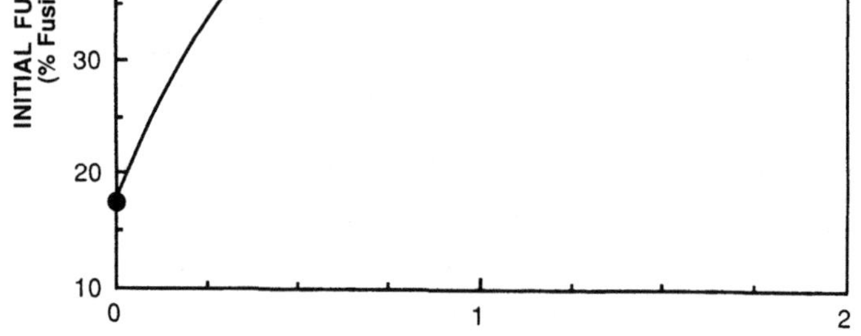


Figure 3. The initial rate of fusion of $SIV_{mac}$ with cardiolipin liposomes as a function of the calcium ion concentration. Reprinted with permission from Larsen et al., 1990a.

NBD-phosphatidyl-ethanolamine (PE) and rhodamine-PE from labeled liposomes into viral membranes during fusion (Hoekstra et al., 1984; Stegmann et al., 1986).

Figure 2 shows the time-course of fusion of $SIV_{mac}$ with liposomes of varying composition, measured as the dequenching of R-18 incorporated into the virus membrane (Larsen et al., 1989; 1990a). To ascertain whether the probe exchanges between membranes in the absence of fusion, labeled $SIV_{mac}$ was incubated with unlabeled $SIV_{mac}$; no increase in fluorescence was observed, indicating that the probe remains stably in the viral membrane.

The fastest fusion was obtained with pure cardiolipin vesicles. The rate and extent of fusion were lower with cardiolipin: DOPC (3:7) liposomes. With pure phosphatidylserine, fusion was slower than with cardiolipin-containing vesicles. The fusion of $SIV_{mac}$ with pure DOPC vesicles was very slow. Thus, the rate and extent of fusion of $SIV_{mac}$ with liposomes depend strongly on the phospholipid composition of the latter. These results also indicate that the virus can fuse with membranes that do not have the CD4 receptor. The rate of fusion with cardiolipin liposomes increased significantly in the presence of calcium in the medium (Figure 3). Calcium up to 5 mM had no effect on the fusion of the virus with DOPC liposomes, indicating that calcium does not activate the virus to fuse rapidly with a

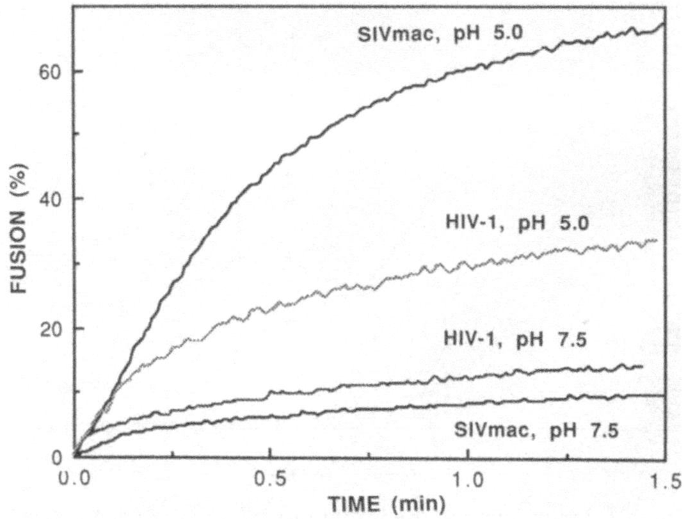


Figure 4. The effect of mildly acidic pH on the fusion of $SIV_{mac}$ or HIV-1 with cardiolipin liposomes. HIV-$1_{LAV}$ was cultured in CEM cells, concentrated by centrifugation, and purified by sucrose density gradient centrifugation. Reproduced with permission from Larsen et al., 1990a.

membrane that it does not fuse with appreciably in the absence of the divalent cation. The enhancement of the rate of fusion of $SIV_{mac}$ by calcium is in contrast to the calcium-independence of the fusion of Sendai and influenza viruses (Stegmann et al., 1985; D. Hoekstra, personal communication).

Although $SIV_{mac}$ exhibited fusion activity at neutral pH, this activity was modulated by low pH. With pure cardiolipin liposomes, lowering the pH to 5 drastically increased the initial rate and extent of fusion over the rate and extent obtained at neutral pH (Figure 4). Similar results were obtained with cardiolipin/DOPC (3:7) or phosphatidylserine liposomes as target membranes. The plot of the initial rates of fusion as a function of pH is shown in Figure 5.

With DOPC liposomes, the initial rate and extent of fusion were very low, during the time of observation (about 5 min), even at acidic pH. With cardiolipin/ DOPC liposomes, the the initial rate of fusion began to increase sharply as the pH was lowered to 5.5. With phosphatidylserine, the initial rate of fusion increased abruptly below pH 5. The inset in Figure 4 shows the pH-dependence of fusion on an expanded scale. For comparison, the pH dependence of the fusion of influenza

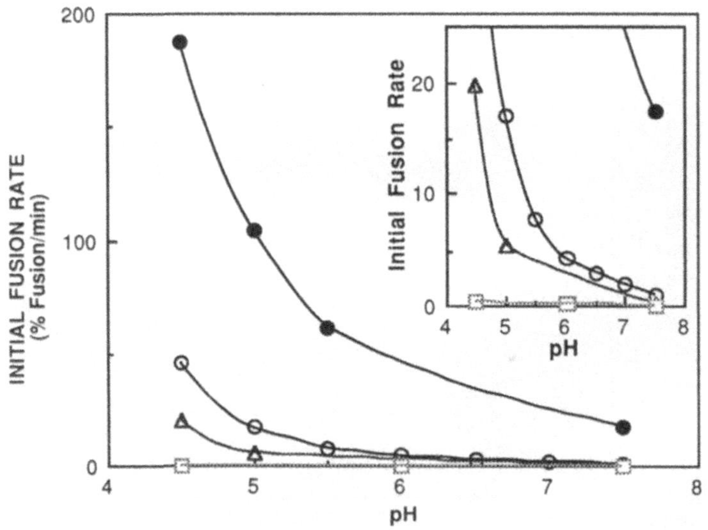


Figure 5. The pH-dependence of the initial rate of fusion of $SIV_{mac}$ with liposomes of various phospholipid composition. Cardiolipin: Closed circles. Cardiolipin/DOPC: Open circles. Phosphatidylserine: Triangles. DOPC: Squares. Reproduced with permission from Larsen et al., 1990a.

virus and Sendai virus with various liposomes is given in Figures 6 and 7, respectively.

Gangliosides act as receptors for influenza and Sendai viruses (Stegmann et al., 1989b; Haywood and Boyer, 1984, 1985). We therefore investigated the fusion of $SIV_{mac}$ with DOPC liposomes containing 15 mol% disialoganglioside ($G_{D1a}$). The rate and the extent of fusion were similar to that obtained with phosphatidylserine liposomes, and low pH did not significantly alter the fusion activity (Larsen et al., 1990a).

To investigate whether the fusion of $SIV_{mac}$ with cardiolipin liposomes involved the envelope glycoprotein gp120, we studied the effect of recombinant soluble CD4 (rsCD4). Soluble CD4 has been shown to bind to gp120 with an affinity similar to that of membrane-bound CD4 and to inhibit infection of cultured cells by HIV (Fisher et al., 1988; Hussey et al., 1988; Byrn et al., 1989). The rate and extent of fusion of $SIV_{mac}$ with cardiolipin liposomes were both reduced in the presence of 10-100 μg/ml rsCD4 (Figure 8). A similar inhibition of fusion by rsCD4 was observed in the presence of 0.5 mM $Ca^{2+}$ (Figure 9). We should note, however, that the concentrations of rsCD4 necessary for inhibition were rather high. The infectivity of primary isolates of HIV-1 also does not appear to be inhibitable with low concentrations of rsCD4 (Daar et al., 1990). However, control experiments utilizing bovine serum albumin instead of rsCD4 resulted in a similar inhibition. These observations indicate that blocking the CD4-binding site of gp120 is not particularly inhibitory to fusion, and that the interaction of $SIV_{mac}$ with liposomes may be independent of gp120. Our observation that trypsin treatment of HIV-1 inhibits its fusion with liposomes (see below) implicates the involvement of

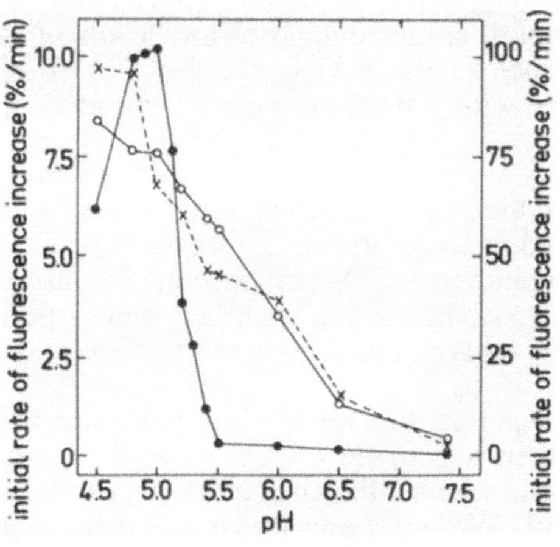


Figure 6. The pH-dependence of the initial rate of fusion of influenza virus with liposomes. Open circles and x: Cardiolipin liposomes (right hand scale). Filled circles: Liposomes composed of dioleoylphosphatidylcholine/dioleoylphosphatidylethanolamine/ gangliosides (6:3:1). Experiments were performed with either the R-18 assay (circles), or the NBD/rhodamine resonance energy transfer assay (x). Reproduced with permission from Stegmann et al., 1986.

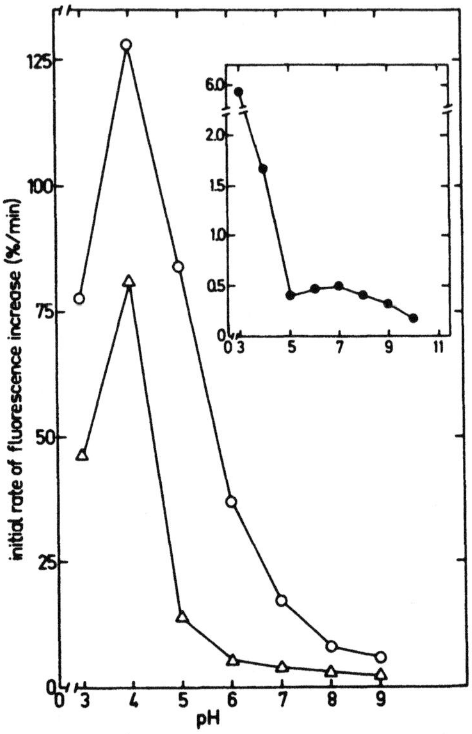


Figure 7. The pH-dependence of the initial rate of fusion of Sendai virus with liposomes composed of cardiolipin (open circles), phosphatidylserine (open triangles) or dioleoylphosphatidylcholine/ dioleoylphosphatidyl-ethanolamine/cholesterol/$G_{D1a}$ (7:3:6.6:1.1) (filled circles, insert). Reproduced with permission from Klappe et al., 1986.

the envelope protein in the fusion reaction. Thus, it may be segments of gp41 that mediate the binding and fusion of the virus with liposomes. Alternatively, parts of gp120 not involved in binding to CD4 may mediate the fusion process (Freed et al., 1991). The rsCD4 may also be playing the role of a non-specific protein that acts as a spacer between the virus and the liposome membrane.

While $SIV_{mac}$ can fuse with membranes containing negatively charged lipids, it was also of interest from the point of view of the mechanisms of this interaction, to investigate whether the virus can fuse with liposomes composed of a positively charged lipid. We had shown previously that liposomes containing N[1-(2,3-dioleyloxypropyl]-N,N,N-trimethylammonium (DOTMA) fuse with certain negatively charged liposomes (Düzgüneş et al., 1989). When $SIV_{mac}$ and DOTMA/cholesterol liposomes were mixed, rapid fusion ensued. Fusion was independent of calcium, was not enhanced by low pH, and was not inhibitable by rsCD4 (C. Larsen, D. Alford, L. Young, T. McGraw, N. Düzgüneş, unpublished results; Düzgünes et al., 1991b). It is likely that DOTMA interacts directly with a

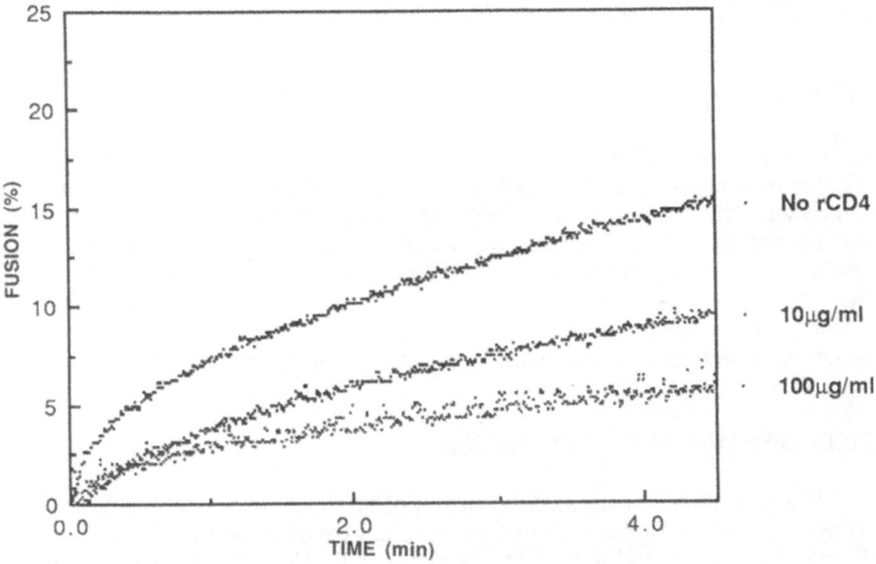


Figure 8. The effect of recombinant soluble CD4 on the fusion of $SIV_{mac}$ with cardiolipin liposomes.

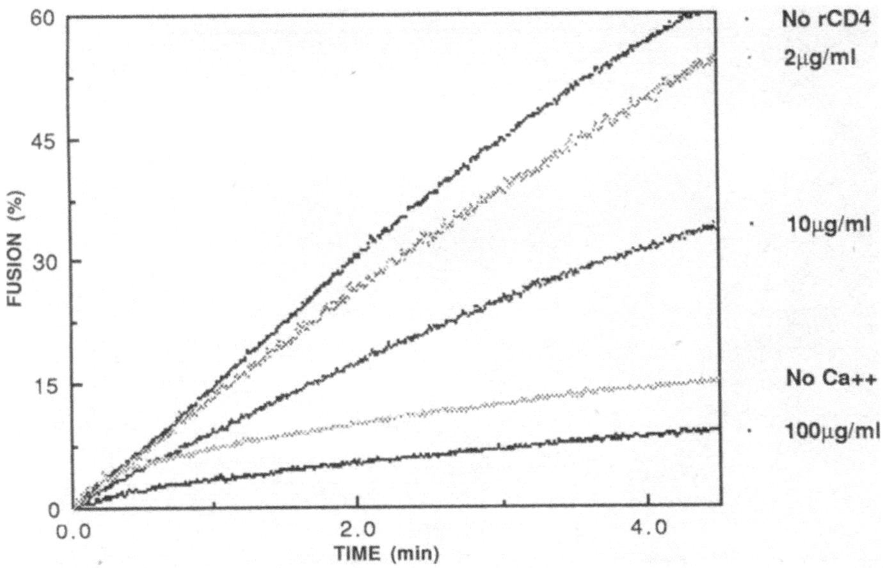


Figure 9. The effect of recombinant soluble CD4 on the fusion of $SIV_{mac}$ with cardiolipin liposomes in the presence of 0.5 mM $Ca^{2+}$.

lipid component of the $SIV_{mac}$ membrane, but it cannot be ruled out that it also interacts with acidic amino acids on the viral proteins.

## FUSION OF $SIV_{mac}$ WITH ERYTHROCYTE GHOSTS

Erythrocyte ghost membranes have been used as a convenient biological target membrane to study the fusion activity of various lipid-enveloped viruses, including influenza virus (Stegmann et al., 1986) and Sendai virus (Hoekstra and Klappe, 1986). Fusion of $SIV_{mac}$ with erythrocyte ghosts was very slow at neutral pH, but was enhanced at acidic pH (Figure 10). The rate of fusion increased further when calcium was also present in the low pH medium, but not at neutral pH (Düzgünes et al., 1991b). Thus, the enhancement of $SIV_{mac}$ fusion activity by calcium or low pH is not restricted to its fusion with liposomes as target membranes.

## FUSION OF HIV-1 WITH LIPOSOMES

We have also investigated the fusion of HIV-1 with liposomes (Larsen et al., 1990a,b; 1991). Figure 4 shows the time course of fusion of HIV-1 with cardiolipin liposomes, and the effect of mildly acidic pH. The rate and extent of

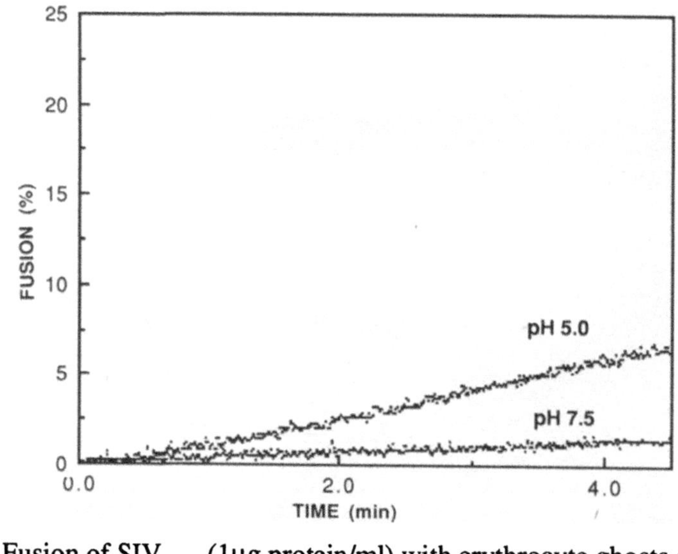


Figure 10. Fusion of $SIV_{mac}$ (1μg protein/ml) with erythrocyte ghosts at 37° C, and the effect of acidic pH. Erythrocyte ghost membranes were formed by hypotonic lysis and re-sealing of fresh human red blood cells (Dodge et al., 1963), and stored in the NACl/TES/citrate buffer.

fusion were comparable to that of $SIV_{mac}$. Lowering the pH to 5 enhanced both the initial rate and extent of fusion, but the extent of fusion was lower than that obtained with $SIV_{mac}$. The lipid composition of the liposomes affected the rate of fusion at neutral pH, with the rate decreasing in the order cardiolipin > cardiolipin/DOPC (3:7) > phosphatidylinositol > phosphatidic acid > phosphatidylserine > DOPC (Larsen et al., 1991). The presence of 33 mol% cholesterol in phosphatidylserine liposomes did not alter the rate of fusion, compared to pure phosphatidylserine liposomes. Low pH enhanced the rate of fusion of HIV-1 with liposomes containing anionic phospholipids, but not with DOPC liposomes. The negligible levels of fusion obtained with DOPC liposomes renders this lipid a good candidate for use in the reconstitution of the complete (trans-membrane) CD4 molecule in liposomes, to study the role of this protein in the fusion of $SIV_{mac}$ or HIV-1 with target membranes.

The physiological significance of the enhancement of HIV-1 and $SIV_{mac}$ fusion at mildly acidic pH is not known at present. Clearly, low pH is not required to initiate fusion. It is, however possible that endocytosed virus particles may have enhanced fusion activity when the pH in the endosome is lowered (Ohkuma and Poole, 1980; Tycko and Maxfield, 1982). This hypothesis is currently under investigation, using cultured T-cells and adherent cells stably transfected with CD4 as targets for the virus.

Similar to $SIV_{mac}$, HIV-1 could also fuse with cationic DOTMA-containing liposomes (Konopka et al., 1990; Figure 11).

When HIV was treated with trypsin, the fusion activity of HIV-1 was diminished, indicating that the envelope protein of the virus is crucial to the process

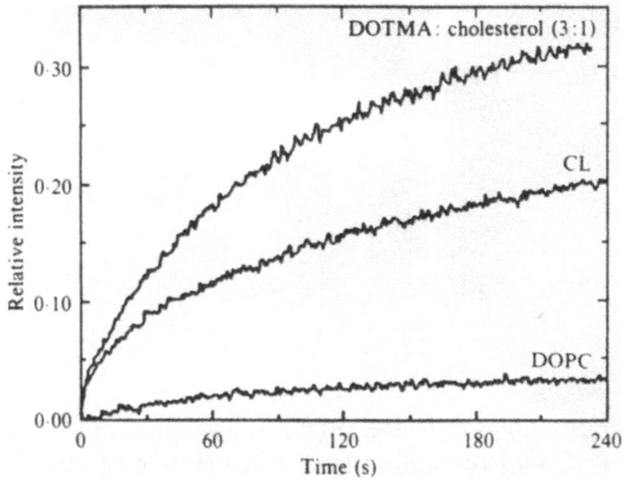


Figure 11. Kinetics of fusion of HIV-1 (1 µg protein/ml) with liposomes (50 µM lipid) composed of N[1-(2,3-dioleyloxy-propyl]-N,N,N-trimethylammonium (DOTMA)/cholesterol, cardiolipin, or dioleoylphosphatidylcholine (DOPC). Reproduced with permission from Konopka et al., 1990.

of fusion. An analysis of the extents of fusion at different virus-to-liposome ratios indicated that the fusion products of HIV-1 and cardiolipin liposomes consisted of one virus and any number of liposomes (Larsen et al., 1991). The ability of the fusion products of HIV-1 and liposomes to fuse with plain liposomes was greatly diminished. It was, therefore, of interest to investigate the effect of liposomes on the infectivity of HIV-1.

## MODULATION OF HIV-1 INFECTIVITY BY LIPOSOMES

When HIV-1 was pre-incubated with cardiolipin liposomes for 2 h at 37° C, and then added to CD4+ human lymphoblastoid A3.01 cells, the infectivity of the virus was markedly diminished (Konopka et al., 1990; Figure 12). The infectivity was measured by the production of viral gag (p24) protein in culture supernatants. The inhibition of infectivity by cardiolipin was concentration-dependent. When cells were exposed to virus in the presence of cardiolipin liposomes a similar inhibition of infectivity was observed, and this inhibition was specific for cardiolipin (Konopka et al., 1991b). The inhibitory effect of cardiolipin liposomes was more pronounced when the HTLV-$III_B$ strain of HIV-1 was tested (K. Konopka and N. Düzgüneş, unpublished data).

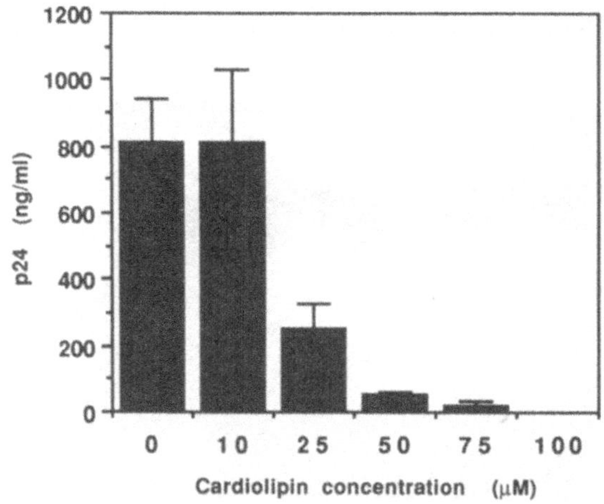


Figure 12. The effect of pre-incubation of HIV-$1_{LAV}$ with cardiolipin liposomes (2 h at 37° C) on virus replication in A3.01 cells, measured as the production of p24 in the culture supernatant after a 10 day incubation. The pre-treated virus was incubated with the cells for 2 h , the unbound virus was washed, and the cells were incubated at 37° C in RPMI 1640/10% heat-inactivated fetal bovine serum. The virus stock was propagated in A3.01 cells. The p24 value at 100 µM is 0.7 ± 0.6. Data from Konopka et al., 1990.

This inhibition may be the result of decreased binding of the virus to cells in the presence of cardiolipin, or the decreased fusogenic capacity of the HIV-1-cardiolipin fusion product. Fusion of cardiolipin liposomes with the viral membrane may render the envelope glycoproteins incapable of maintaining a proper conformation to sustain further fusion as the proteins diffuse laterally into the liposomal membrane. Previous studies on the ability of phosphatidylserine liposomes to block the infection of Vero cells by vesicular stomatitis virus have suggested that this phospholipid may act as a receptor for the virus (Schlegel et al., 1983). Activated peripheral blood mononuclear cells display epitopes recognized by affinity purified anti-cardiolipin antibodies (Misra et al., 1989). It is thus possible that HIV-1 interacts with molecules on the cell surface that resemble cardiolipin; these could be proteins, but they could also be polyphosphoinositides or even cardiolipin.

Completely different results were obtained when HIV-1 was pre-incubated with DOTMA liposomes (Konopka et al., 1990). The infectivity towards A3.01 cells of HIV-1 pre-treated with varying concentrations of DOTMA liposomes is shown in Figure 13. At the optimum lipid concentration, the infectivity was enhanced by about 30-fold. The enhancement of infectivity was even greater when

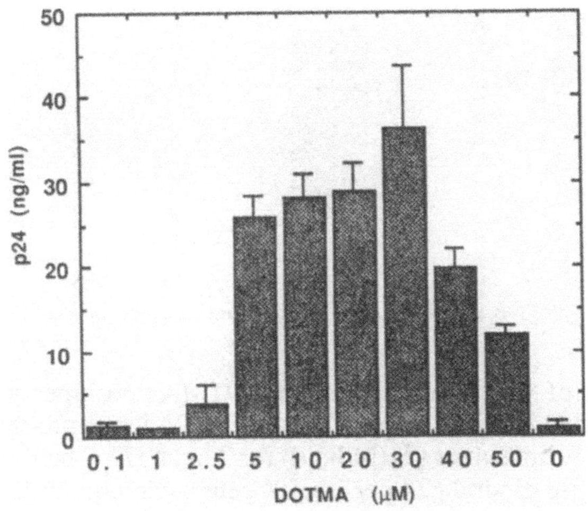


Figure 13. The effect of DOTMA liposomes on HIV-1 replication in A3.01 cells. The virus (0.4 μg p24/ml) was pre-incubated with various concentrations of liposomes (given as μM lipid) for 2 h at 37° C, and then diluted with a cell suspension to give 1-2 ng p24/$10^6$ cells. After virus adsorption for 2 h at 37° C, the cells were washed three times and resuspended at $5x10^5$ cells/ml. Culture supernatants were collected 7 days later and tested for p24 levels by ELISA. Data represent mean values ± S.D. (n = 4). Reproduced with permission from Konopka et al., 1990.

DOTMA liposomes were present during the addition of the virus to A3.01 cells (Figure 14, column 2), amounting to a 90-fold enhancement over the infectivity of untreated HIV-1 (column 4). If the liposomes were added following the initial 2 h inoculation of the cells, only a 2-fold enhancement of infectivity was observed (column 3).

The infectivity of HIV-1 in CD4+ cells can be completely inhibited by anti-Leu3a antibodies that recognize the region on CD4 involved in binding to gp120 (Sattentau et al., 1986; McDougal et al., 1986b; Weber et al., 1989; Konopka et al., 1991a). Similar results were obtained with the A3.01 cells used in our studies. Anti-Leu3a antibodies also inhibited by 93% the infectivity of HIV-1 in the presence of DOTMA liposomes, indicating that DOTMA-mediated enhancement of infectivity takes place primarily via the CD4 receptor (Konopka et al., 1990). Some infection appears to take place independent of CD4, and may be mediated by the

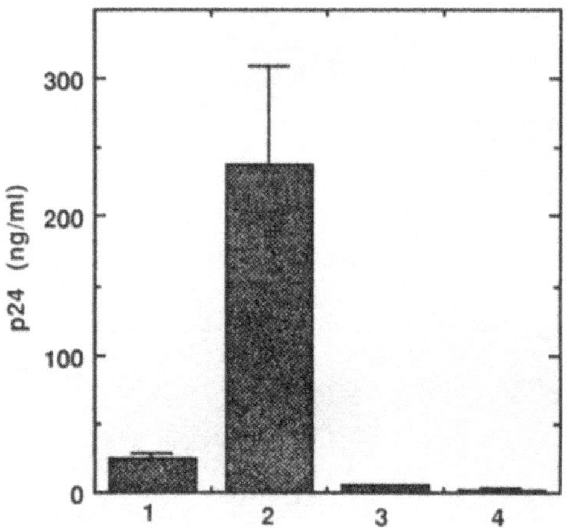


Figure 14. The role of the time of addition of DOTMA liposomes on HIV-1 replication in A3.01 cells. Column 1: HIV-1 was pre-incubated with DOTMA liposomes (5 μM lipid) for 2 hr at 37° C before infection of the cells (1 ng of viral p24 per 1 x $10^6$ cells). Column 2: Cells were incubated with HIV-1 in the presence of 5 μM DOTMA for 2 hr, then washed. Column 3: Cells were infected with HIV-1, then washed and resuspended in medium containing 5 μM DOTMA at 5 x $10^5$ cells/ml. On 3 days post-infection 400 μl samples of the culture supernatant were removed and then fresh medium with 5 μM DOTMA was added to each well. Column 4: Cells were infected with HIV-1 and cultured in the absence of DOTMA. In all cases, p24 antigen was detected in cell culture supernatant on 7 days post-infection. Data represent mean values ± SD (n=6). Reproduced with permission from Konopka et al., 1990.

fusion of the DOTMA-HIV-1 fusion product with the plasma membrane, since DOTMA liposomes can fuse with the plasma membrane of A3.01 cells (L. Stamatatos and N. Düzgünes, unpublished data).

Because of the possibility that DOTMA could mediate the fusion of HIV-1 with the plasma membrane in a CD4-independent manner, we investigated whether the $CD4^-$ human chronic myelogenous leukemia cell line K562 could be infected with HIV-1 pre-treated with DOTMA liposomes. No p24 could be detected in the culture supernatants of these cells incubated with untreated or DOTMA-pre-treated HIV-1. Although a very small amount of HIV-1 proviral DNA (2-10 copies/$10^5$ cells) could be detected in cells incubated for 3 days following a 2 h incubation with treated or untreated HIV-1, it was not possible to distinguish between newly synthesized DNA and contaminating viral DNA from infected cells in the viral inoculum. Furthermore, no DNA was detectable 10 days following inoculation. Thus pre-treatment with DOTMA liposomes did not appear to induce or enhance HIV-1 infectivity in these cells (Konopka et al., 1990). This observation is in contrast with the ability of lipofectin, which contains DOTMA and dioleoylphosphatidylethanolamine, to mediate the infection of non-permissive cells by murine leukemia virus (Innes et al., 1990). Again in contrast to our observations in CD4+ cells, lipofectin did not enhance the infectivity of murine leukemia virus in permissive cells.

## LIPOSOMES AS TARGET MEMBRANES FOR VIRUS FUSION

The use of liposomes as target membranes for viruses have been useful in understanding the molecular requirements for viral membrane fusion. The use of liposomes was crucial for the demonstration that Semliki Forest virus requires cholesterol in the target membrane (White and Helenius, 1980; Kielian and Helenius, 1984), and that acidic pH is required for the induction of fusion between Semliki Forest virus and target membranes (Helenius et al., 1980). The amphiphilic properties of influenza virus hemagglutinin were demonstrated by studying the interaction of the protein with liposomes (Doms et al., 1985), and the demonstration that the conformational change of the hemagglutinin per se is not sufficient for fusion was made using liposomes as target membranes (Stegmann et al., 1985). Liposomes, as target membranes composed only of phospholipids, were also essential for the demonstration of the penetration of segments of viral envelope proteins into the target membrane during fusion (Novick and Hoekstra, 1988; Harter et al., 1989). Fusion of viruses with liposomes provided a well defined experimental system for the determination of viral adhesion and fusion rate constants (Nir et al., 1986a,b).

As pointed out above, the physiological significance of our observations with liposomes as target membranes remains to be established. That the CD4 receptor is not necessary for the fusion of HIV-1 or $SIV_{mac}$ is one of the significant results of our studies. The enhancement of fusion at low pH is a novel observation for immunodeficiency viruses, and needs to be investigated further with cellular membranes as targets. In contrast to our results with liposomes, Sinangil et al. (1988) found, using the R-18 assay, that the fusion of HIV-1 with CD4+ CEM cells was optimal at pH 7, but also occured at pH 5. The effect of calcium is also unusual, since the fusion activity of other lipid enveloped viruses do not appear to be influenced by this cation.

## THERAPEUTIC IMPLICATIONS OF THE MODULATION OF HIV INFECTIVITY BY LIPOSOMES

The inhibition of HIV-1 infectivity with cardiolipin liposomes raises the possibility that such liposomes could be used for therapeutic purposes. Such liposomes may not only prevent entry of virions into target cells, but may also encounter intact virions in vacuoles of mononuclear phagocytes (Gartner et al., 1986; Nicholson et al., 1986; Gendelman et al., 1988) following endocytosis, and render those virus particles incapable of infecting other cells if they are released from the phagocyte. Encapsulation of anti-sense oligodeoxynucleotides or reverse transcriptase inhibitors, which would enter the virus interior via fusion, could further inhibit HIV infectivity. These inhibitors could thus be carried into the cytoplasm by any virus-liposome fusion product that may still be fusogenic towards the cell membrane. In preliminary experiments we have observed that infectivity was completely inhibited when an anti-rev phosphorothioate oligodeoxynucleotide was encapsulated in cardiolipin liposomes (K. Konopka and N. Düzgüneş, unpublished data). Encapsulated proteases and nucleases in liposomes may degrade the virus and its genome upon fusion with the viral membrane, and confer further protection against HIV infection. The reconstitution of CD4 in the liposome membrane or the covalent coupling of rsCD4 to the liposome surface may render the liposomes highly specific for binding and fusion with HIV-1. Since cardiolipin-containing liposomes for the delivery of doxorubicin in cancer chemotherapy are in phase II clinical trials (Treat et al., 1989), it is not unlikely that similar liposomes may be useful for the chemotherapy of HIV infections.

The enhancement of the infectivity of HIV-1 by DOTMA liposomes raises questions aboout the use of such liposomes for the delivery of anti-sense nucleic acids or gene-regulatory proteins (Felgner et al., 1987; Debs et al., 1990) into HIV-infected cells. Nevertheless, the enhancement being minimal when DOTMA liposomes are added following the inoculation of the cells with HIV-1 (FIgure 14, column 3), in contrast to being present during infection (Figure 14, column 2), may still allow the use of such liposomes if the inhibitory effects of the delivered compounds outweigh the enhancement of replication.

## ACKNOWLEDGMENTS

We thank Dr. Daniel Stites for the use of his P-2 laboratory, and Dr. Demetrios Papahadjopoulos for the use of his laboratory facilities and helpful discussions. This work was supported by Grants AI25534 (N.D.) and AI26471 (T.P.M.), and Fellowship AI08117 (C.E.L.) from the National Institute of Allergy and Infectious Diseases, Grant HL42105 (B.D.) from the National Heart, Lung, and Blood Institute, Grant 86SF017 from the State of California Universitywide AIDS Research Program (N.D), a grant from Liposome Technology, Inc. (N.D. & K.K.), U.S.-Israel Binational Science Foundation Grant 86-00010 (N.D. & S.N.) and Fellowship PF-3394 from the American Cancer Society (C.E.L.).

REFERENCES

Barré-Sinoussi, F., Chermann, J. C., Rey, F., Nugeyre, M. T., Chamaret, S., Gruest, J., Dauguet, C., Axler-Blin, C., Brun-Vézinet, F., Rouzioux, C., Rozenbaum, W., & Montagnier, L., 1983, Isolation of a T-lymphotropic retrovirus from a patient at risk for acquired immune deficiency syndrome (AIDS), Science 220:868-871.

Bosch, M. L., Earl, P. L., Fargnoli, K., Picciafuoco, S., Giombini, F., Wong-Staal, F., and Franchini, G., 1989, Identification of the fusion peptide of primate immunodeficiency viruses, Science 244:694-697.

Byrn, R. A., Sekigawa, I., Chamow, S. M., Johnson, J. S., Gregory, T. J., Capon, D. J. and Groopman, J. E., 1989, Characterization of in vitro inhibition of human immunodeficiency virus by purified recombinant CD4, J. Virol. 63:4370-4375.

Chakrabarti, L., Guyader, M., Alizon, M., Daniel, M. D., Desrosiers, R. C., Tiollais, P., and Sonigo, P., 1987, Sequence of simian immunodeficiency virus from macaque and its relationship to other human and simian retroviruses, Nature 328:543-547.

Cheng-Mayer, C., Rutka, J. T., Rosenblum, M. L., McHugh, T., Stites, D. P. and Levy, J. A., 1987, Human immunodeficiency virus can productively infect cultured human glial cells, Proc. Natl. Acad. Sci. USA 84:3526-3530.

Clapham, P. R., Weber, J. N., Whitby, D., McIntosh, K., Dalgleish, A. G., Maddon, P. J., Deen, K. C., Sweet, R. W. and Weiss, R. A., 1989, Soluble CD4 blocks the infectivity of diverse strains of HIV and SIV for T cells and monocytes but not for brain and muscle cells, Nature 337:368-370.

Clavel, F., Guétard, D., Brun-Vézinet, F., Chamaret, S., Rey, M.-A., Santos-Ferreira, M. O., Laurent, A. G., Dauguet, C., Katlama, C., Rouzioux, C., Klatzmann, D., Champalimaud, J. L., and Montagnier, L., 1986, Isolation of a new human retrovirus from West African patients with AIDS, Science 233:343-346.

Daar, E. S., Li, X. L., Moudgil, T., Ho, D. D., 1990, High concentrations of recombinant soluble CD4 are required to neutralize primary human immunodeficiency virus type 1 isolates, Proc. Natl. Acad. Sci. USA 87:6574-6578.

Dalgleish, A. G., Beverly, P. C. L., Clapham, P. R., Crawford, D. H., Greaves, M. F., and Weiss, R. A., 1984, The CD4 (T-4) antigen is an essential component of the receptor for the AIDS retrovirus, Nature 312:763-766.

Debs, R. J., Freedman, L. P., Edmunds, S., Gaensler, K. L., Düzgünes, N. and Yamamoto, K. R., 1990, Regulation of gene expression in vivo by liposome-mediated delivery of a purified transcription factor, J. Biol. Chem. 265:10189-10192.

Desrosiers, R. C., 1988, Simian immunodeficiency viruses. Annu. Rev. Microbiol. 42:607-625.

DeVita, V.T., Hellman, S. and Rosenberg, S. A., 1988, "AIDS: Etiology, Diagnosis, Treatment, and Prevention," Lippincott, Philadelphia.

Dodge, J. T., Mitchell, C., and Hanahan, D. J., 1963, The preparation and chemical characteristics of hemoglobin-free ghosts of human erythrocytes, Arch. Biochem. Biophys. 100:119-130.

Doms, R. W., Helenius, A. and White, J., 1985, Membrane fusion activity of the influenza virus hemagglutinin: The low pH-induced conformational change, J. Biol. Chem. 260:2973-2981.
Düzgünes, N., 1985, Membrane fusion, Subcell. Biochem. 11:195-286.
Düzgünes, N. and Shavnin, S. A., 1991, N-terminal peptides of viral fusion proteins: Interaction with phospholipid vesicles, Biochim. Biophys. Acta (submitted).
Düzgünes, N., Wilschut, J., Fraley, R. and Papahadjopoulos, D., 1981, Studies on the mechanism of membrane fusion: Role of head-group composition in calcium- and magnesium-induced fusion of mixed phospholipid vesicles, Biochim. Biophys. Acta 642:182-195.
Düzgünes, N., Goldstein, J. A., Friend, D. S. and Felgner, P. L., 1989, Fusion of liposomes containing a novel cationic lipid, N[1-(2,3-dioleyloxypropyl]-N,N,N-trimethyl-ammonium: Induction by multivalent anions and asymmetric fusion with acidic phospholipid vesicles, Biochemistry 28:9179-9184.
Düzgünes, N., Pedroso de Lima, M. C., Stamatatos, L., Flasher, D., Alford, D., Friend D. S. and Nir, S., 1991a, Fusion activity and inactivation of influenza virus: Kinetics of low-pH induced fusion with cultured cells, J. Gen. Virol. (submitted).
Düzgünes, N., Larsen, C. E., Stamatatos, L. and Konopka, K., 1991b, Fusion of immunodeficiency viruses with liposomes and cells: Inhibition of HIV-1 infectivity by liposomes, in: "Membrane Interactions of Human Immunodeficiency Virus," R. C. Aloia, C. C. Curtain and L. M. Gordon, eds., Wiley -Liss, New York (in preparation).
Felgner, P. L., Gadek, T. R., Holm, M., Roman, R., Chan, H. W., Wenz, M., Northrop, J. P., Ringold, G. M., and Danielsen, M., 1987, Lipofectin: a highly efficient, lipid-mediated DNA-transfection procedure, Proc. Natl. Acad. Sci. U.S.A. 84:7413-7417.
Fisher, R. A., Bertonis, J. M., Meier, W., Johnson, V. A., Costopoulos, D. S., Liu, T., Tizard, R., Walker, B. D., Hirsch, M. S., Schooley, R. T. and Flavell, R. A., 1988, HIV infection is blocked in vitro by recombinant soluble CD4, Nature 331:76-78.
Freed, E. O., Myers, D. J. and Risser, R., 1991, Identification of the principal neutralizing determinant of human immunodeficiency virus type 1 as a fusion domain, J. Virol. 65:190-194.
Gallaher, W. R., 1987, Detection of a fusion peptide sequence in the transmembrane protein of human immunodeficiency virus, Cell 50:327-328.
Gallo, R. C., Salahuddin, S. Z., Popovic, M., Shearer, G. M., Kaplan, M., Haynes, B. F., Palker, T. J., Redfield, R., Oleske, J., Saafai, B., White, G., Foster, P., and Markham, P. D., 1984, Frequent detection and isolation of cytopathic retroviruses (HTLV-III) from patients with AIDS, Science 224:500-503.
Gartner, S., Markovits, P., Makovitz, D. M., Kaplan, M. H., Gallo, R. C., Popovic, M., 1986, The role of mononuclear phagocytes in HTLV-III/LAV infection, Science 233:215-219.
Gendelman, H. E., Orenstein, J. M., Martin, M. A., Ferrua, C., Mitra, R., Phipps, T., Wahl, L. A., Lane, H. C., Fauci, A. S., Burke, D. S., Skillman, D. and Meltzer, M. S., 1998, Efficient isolation and propagation

of human immunodeficiency virus on recombinant colony-stimulating factor 1-treated monocytes, J. Exp. Med. 167:1428-1441.

Gonzalez-Scarano, F., Waxham, M. N., Ross, A. M. and Hoxie, J. A., 1987, Sequence similarities between human immunodeficiency virus gp41 and paramyxovirus fusion proteins, AIDS Res. Human Retrovir. 3:245-252.

Gowda, S. D., Stein, B. S., Mohagheghpour, N., Benike, C. J., and Engleman, E. G., 1989, Evidence that T cell activation is required for HIV-1 entry in $CD4^+$ lymphocytes, J. Immunol. 142:773-780.

Harouse, J. M., Kunsch, C., Hartle, H. T., Laughlin, M. A., Hoxie, J. A., Wigdahl, B., and Gonzalez-Scarano, F., 1989, CD4-independent infection of human neural cells by human immunodeficiency virus type 1, J. Virol. 63:2527-2533.

Harter, C., James, P., Bächi, T., Semenza, G. and Brunner, J., 1989, Hydrophobic binding of the ectodomain of influenza hemagglutinin to membranes occurs through the "fusion peptide." J. Biol. Chem. 264:6459-6464.

Haywood, A. M. and Boyer, B. P., 1984, Effect of lipid composition upon fusion of liposomes with Sendai virus membranes, Biochemistry 23:4161-4166.

Haywood, A. M. and Boyer, B. P., 1985, Fusion of influenza virus membranes with liposomes at pH 7.5, Proc. Natl. Acad. Sci. USA 82:4611-4615.

Helenius, A., Kartenbeck, J., Simons, K. and Fries, E., 1980, On the entry of Semliki Forest virus into BHK-21 cells, J. Cell Biol. 84:404-420.

Hoekstra, D. and Klappe, K., 1986, Sendai virus-erythrocyte membrane interaction: Quantitative and kinetic analysis of viral binding, dissociation and fusion, J. Virol. 58:87-95.

Hoekstra, D., and Kok, J. W., 1989, Entry mechanisms of enveloped viruses. Implications for fusion of intracellular membranes, Biosci. Rep. 9:273-305.

Hoekstra, D., de Boer, T., Klappe, K., and Wilschut, J., 1984, Fluorescence method for measuring the kinetics of fusion between biological membranes, Biochemistry 23:5675-5681.

Hoekstra, D., Klappe, K., de Boer, T., and Wilschut, J., 1985, Characterization of the fusogenic properties of Sendai virus: Kinetics of fusion with erythrocyte membranes, Biochemistry 24:4739-4745.

Hussey, R. E., Richardson, N. E., Kowalski, M., Brown, N. R., Chang, H. C., Siliciano, R. F., Dorfman, T., Walker, B., Sodroski, J. and Reinherz, E. L., 1988, A soluble CD4 protein selectively inhibits HIV replication and syncytium formation, Nature 331:788-81.

Innes, C. L., Smith, P. B., Langenbach, R., Tindall, K. R. and Boone, L. R., 1990, Cationic liposomes (Lipofectin) mediate retroviral infection in the absence of specific receptors, J. Virol. 64:957-961.

Kannagi, M., Yetz, J. M., and Letvin, N. L, 1985, In vitro growth characteristics of simian T-lymphotropic virus type III, Proc. Natl. Acad. Sci. USA 82:7053-7057.

Kielian, M., and Helenius, A., 1984, Role of cholesterol in fusion of Semliki Forest virus with membranes, J. Virol. 52:281-283.

Klappe, K., Wilschut, J., Nir, S. and Hoekstra, D., 1986, Parameters affecting fusion between Sendai virus and liposomes. Role of viral proteins, liposome composition, and pH, Biochemistry 25:8252-8260.

Klatzmann, D., Champagne, E., Chamaret, S., Gruest, J., Guétard, D., Hercend, T., Gluckman, J.-C., and Montagnier, L., 1984, T-lymphocyte T4 molecule behaves as the receptor for human retrovirus LAV, Nature 312:767-768.
Koenig, S., Hirsch, V. M., Olmsted, R. A., Powell, D., Maury, W., Rabson, A., Fauci, A. S., Purcell, R. H., and Johnson, P. R., 1989, Selective infection of human CD4+ cells by simian immunodeficiency virus: Productive infection associated with envelope glycoprotein-induced fusion, Proc. Natl. Acad. Sci. USA 86:2443-2447.
Konopka, K., Davis, B. R., Larsen, C. E., Alford, D. R., Debs, R. J. and Düzgünes, N., 1990, Liposomes modulate human immunodeficiency virus infectivity, J. Gen. Virol. 71:2899-2907.
Konopka, K., Davis, B. R. and Düzgünes, N., 1991a, HIV-1 infection of a non-CD4-expressing variant of HUT-78 cells: Lack of inhibition by Leu3a antibodies and enhancement by cationic DOTMA liposomes, in: "Mechanisms and Specificity of HIV Entry into Host Cells," N. Düzgünes, ed., Plenum Press, New York (this volume).
Konopka, K., Davis, B. R., Larsen, E. and Düzgünes, N., 1991b, Cardiolipin liposomes specifically inhibit HIV-1 infectivity, Antiviral Res. Suppl. I: 68.
Kowalski, M., Potz, J., Basiripour, L., Dorfman, T., Goh, W. C., Terwilliger, E., Dayton, A., Rosen, C., Haseltine, W., and Sodroski, J., 1987, Functional regions of the envelope glycoprotein of human immunodeficiency virus type 1, Science 237:1351-1355.
Larsen, C., Alford, D., Young, L., Lee, K-D., McGraw, T. and Düzgünes, N., 1989, Fusion of simian immunodeficiency virus ($SIV_{mac}$) with liposomes and erythrocyte ghosts, Abstr. V International Conference on AIDS (p. 634), Montreal, Canada.
Larsen, C. E., Alford, D. R., Young, L. J. T., McGraw, T. P. and Düzgünes, N., 1990a, Fusion of simian immunodeficiency virus ($SIV_{mac}$) with liposomes and erythrocyte ghosts: Effect of liposome composition, low pH and calcium, J. Gen. Virol. 71:1947-1955.
Larsen, C. E., Alford, D. R., Jennings, M., Debs, R. and Düzgünes, N., 1990b, Fusion of HIV-1 with liposomes. Abstr. VI International Conference on AIDS (Vol. 2, p. 147), San Francisco, CA.
Larsen, C. E., Alford, D. R., Nir, S., Jennings, M., Lee, K.-D. and Düzgünes, N., 1991, Fusion of HIV-1 with liposomes and erythrocyte ghosts, Biophys. J. 59:131a.
Letvin, N. L., Daniel, M. D., Sehgal, P. K., Desrosiers, R. C., Hunt, R. D., Waldron, L. M., MacKey, J. J., Schmidt, D. K., Chalifoux, L. V., and King, N. W., 1985, Induction of AIDS-like disease in macaque monkeys with T-cell tropic retrovirus STLV-III, Science 230:71-73.
Levy, J. A., Hoffman, A. D., Kramer, S. M., Landis, J. A., Shimabukuro, J. M., and Oshiro, L. S., Isolation of lymphocytopathic retroviruses from San Francisco patients with AIDS, 1984, Science 225:840-842.
Lifson, J. D., Feinberg, M. B., Reyes, G. R., Rabin, L., Banapour, B., Chakrabarti, S., Moss, B., Wong-Staal, F., Steimer, K. S., and Engleman, E. G., 1986, Induction of CD4-dependent cell fusion by the HTLV-III/LAV envelope glycoprotein, Nature 323:725-728.

Loyter, A., Citovsky, V., and Blumenthal, R., 1988, The use of fluorescence dequenching measurements to follow viral membrane fusion events, Methods Biochem. Analysis 33:129-164.

Maddon, P. J., Dalgleish, A. G., McDougal, J. S., Clapham, P. R., Weiss, R. A., and Axel, R., 1986, The T4 gene encodes the AIDS virus receptor and is expressed in the immune system and the brain, Cell 47:333-348.

Maddon, P. J., McDougal, J. S., Clapham, P. R., Dalgleish, A. G., Jamal, S., Weiss, R. A. and Axel, R., 1988, HIV infection does not require endocytosis of its receptor, CD4, Cell 54:865-874.

Marsh, M., and Dalgleish, A., 1987, How do human immunodeficiency viruses enter cells? Immunol. Today 8:369-371.

McClure, M. O., Marsh, M., and Weiss, R. A., 1988, Human immunodeficiency virus infection of CD4-bearing cells occurs by a pH-independent mechanism. EMBO J. 7:513-518.

McDougal, J. S., Mawle, A., Cort, S. P., Nicholson, J. K. A., Cross, G. D., Scheppler-Campbell, J. A., Hicks, D., and Sligh, J., 1985, Cellular tropism of the human retrovirus HTLV-III/LAV, J. Immunol. 135:3151-3162.

McDougal, J. S., Kennedy, M. S., Sligh, J. M., Cort, S. P., Mawle, A. and Nicholson, J. K. A., 1986a, Binding of HTLV-III/LAV to T4+ T cells by a complex of the 110K viral protein and the T4 molecule, Science 231:382-385.

McDougal, S. J., Nicholson, J. K. A., Cross, G. D., Cort, S. P., Kennedy, M. S. and Mawle, A. C., 1986b, Binding of the human retrovirus HTLV-III/LAV/ARV/HIV to the CD4 (T4) molecule: conformation dependence, epitope mapping, antibody inhibition, and potential for idiotypic mimicry, J. Immunol. 137:2937-2944.

Misra, R., Venables, P. J. W., Plater-Zyberk, C., Watkins, P. F., and Maini, R. N., 1989, Anti-cardiolipin antibodies in infectious mononucleosis react with the membrane of activated lymphocytes, Clin. Exp. Immunol. 75:35-40.

Nicholson, J. K. A., Cross, G. D., Callaway, C. S. and McDougal, J. S., 1986, In vitro infection of human monocytes with human T lymphotropic virus type III/lymphadenopathy-associated virus (HTLV-III/LAV), J. Immunol. 137:323-329.

Nir, S., Stegmann, T. and Wilschut, J., 1986a, Fusion of influenza virus with cardiolipin liposomes at low pH: Mass action analysis of kinetics and fusion, Biochemistry 25:257-266.

Nir, S., Klappe, K. and Hoekstra, D., 1986b, Mass action analysis of kinetics and extent of fusion between Sendai virus and liposomes, Biochemistry 25:8261-8266.

Novick, S. L. and Hoekstra, D., 1988, Membrane penetration of Sendai virus glycoproteins during the early stages of fusion with liposomes as determined by hydrophobic photoaffinity labeling, Proc. Natl. Acad. Sci. USA 85:7433-7437.

Ohki, S., 1988, Surface tension, hydration energy and membrane fusion, in: "Molecular Mechanisms of Membrane Fusion," pp. 123-138, S. Ohki, D. Doyle, T. D. Flanagan, S. K. Hui and E. Mayhew, eds., Plenum Press, New York.

Ohkuma, S. and Poole, B., 1980, Fluorescence probe measurement of the intralysosomal pH in living cells and the perturbation by various agents, Proc. Natl. Acad. Sci. USA, 75:3327-3331.
Ohnishi, S.-I., 1988, Fusion of viral envelopes with cellular membranes, in: "Membrane Fusion in Fertilization, Cellular Transport and Viral Infection," pp. 257-296, N. Düzgünes and F. Bronner, eds., Academic Press, San Diego.
Pauza, C. D., and Price, T. M., 1988, Human immunodeficiency virus infection of T cells and monocytes proceeds via receptor-mediated endocytosis. J. Cell Biol. 107:959-968.
Sattentau, Q., Dalgleish, A. G., Weiss, R. A. and Beverly, P. C. L., 1986, Epitopes of the CD4 antigen and HIV infection, Science 234:1120-1123.
Schlegel, R., Tralka, S. T., Willingham, M. C. and Pastan, I., 1983, Inhibition of VSV binding and infectivity by phosphatidylserine: is phosphatidylserine a VSV-binding site, Cell 32:639-646.
Sinangil, F., Loyter, A., and Volsky, D. J., 1988, Quantitative measurement of fusion between human immunodeficiency virus and cultured cells using membrane fluorescence dequenching, FEBS Lett. 239: 88-92.
Sodroski, J., Goh, W. C., Rosen, C., Campbell, K., and Haseltine, W., 1986, Role of the HTLV-III/LAV envelope in syncytium formation and cytopathicity, Nature 322:470-474.
Stegmann, T., Hoekstra, D., Scherphof, G., and Wilschut, J., 1985, Kinetics of pH-dependent fusion between influenza virus and liposomes, Biochemistry 24:3107-3113.
Stegmann, T., Hoekstra, D., Scherphof, G., and Wilschut, J., 1986, Fusion activity of influenza virus. A comparison between biological and artificial target membrane vesicles, J. Biol. Chem. 261:10966-10969.
Stegmann, T., Doms, R.W. and Helenius, A., 1989a, Protein-mediated membrane fusion, Annu. Rev. Biophys. Chem. 18:187-211.
Stegmann, T., Nir, S., and Wilschut, J., 1989b, Membrane fusion activity of influenza virus. Effects of gangliosides and negatively charged phospholipids in target liposomes, Biochemistry 28:1698-1704.
Stein, B. S., Gowda, S. D., Lifson, J. D., Penhallow, R. C., Bensch, K. G., and Engleman, E. G., 1987, pH-Independent HIV entry into CD4-positive T cells via virus envelope fusion to the plasma membrane, Cell 49:659-668.
Szoka, F., Olson, F., Heath, T., Vail, W., Mayhew, E., and Papahadjopoulos, D., 1980, Preparation of unilamellar liposomes of intermediate size (0.1-0.2 um) by a combination of reverse phase evaporation and extrusion through polycarbonate membranes, Biochim. Biophys. Acta 601:559-571.
Tateno, M., Gonzalez-Scarano, F., and Levy, J. A., 1989, Human immunodeficiency virus can infect CD4-negative human fibroblastoid cells, Proc. Natl. Acad. Sci. USA 86:4287-4290.
Treat, J., Greenspan, A. R. and Rahman, A., 1989, Liposome encapsulated doxorubicin. Preliminary results of phase I and phase II trials, in: "Liposomes in the Therapy of Infectious Diseases and Cancer," pp.353-365, G. Lopez-Berestein and I. J. Fidler, eds., Alan R. Liss, New York.
Tycko, B., and Maxfield, F. R., 1982, Rapid acidification of endocytic vesicles containing $\alpha_2$-macroglobulin, Cell 28:643-652.

Weber, J., Clapham, P., McKeating, J. Stratton, M., Robey, E. & Weiss, R., 1989, Infection of brain cells by diverse human immunodeficiency virus isolates: Role of CD4 as receptor, J. Gen. Virol. 70:2653-2660.

White, J. and Helenius, A., 1980, pH-Dependent fusion between the Semliki Forest virus membrane and liposomes, Proc. Natl. Acad. Sci. USA 77:3273-3277.

White, J., 1990, Viral and cellular membrane fusion proteins, Annu. Rev. Physiol. 52:675-697.

## DISCUSSION

Mills (UCSF): I have a question and a comment. The question is: What is the sequence of addition of the CD4, the virus and the liposomes when you're doing the rsCD4 inhibition experiments? The comment is: that in the absence of a better control, rsCD3 or rsCD8 or something, maybe what you could do is mix the rsCD4 with gp120 before you add it to the virus. Presumably that should at least reduce its inhibitory effect, if it's specific.

Düzgünes: That is an excellent suggestion. To answer your question, the recombinant molecule is added to the virus preparation and then liposomes are introduced.

Mills: Okay, so there's a period of ten or fifteen minutes when the virus is stirred around with the CD4 before you put it into the liposomes?

Düzgünes: Five minutes.

Mills: Okay, that should be enough, I would guess. Because doing it that way, you can basically completely inhibit, or 99.9% inhibit, infectivity. It's curious that you're still getting fusion under those circumstances.

Düzgünes: Yes. It's possible that there is a non-physiological component of fusion with cardiolipin vesicles which Wilschut and collaborators have alluded to with respect to influenza virus fusion with cardiolipin vesicles versus fusion with erythrocyte ghosts. They observe differences in the low-pH inactivation of influenza virus. When they pre-incubate the virus at low pH, its fusion with erythrocyte ghosts, for example, is inhibited, but its fusion with cardiolipin liposomes is not inhibited to the same extent, or it is inhibited only slightly. So there is a difference in that respect, at least in the case of influenza virus.

McDougal (Centers for Disease Control): After you add your R-18 to virus, how do you separate unincorporated R-18 from it? Also, you would predict that this R18-loaded virus would be a promiscuous binder based on your liposome results, right? In other words, it binds to erythrocytes so it probably would bind to lymphocytes. Is there any fundamental difference between the lipid composition of erythrocytes and lymphocytes?

Düzgünes: Regarding your first question as to how the R-18 was separated, in early experiments, we separated them on a Sephadex G-75 column, which was the original procedure used by Hoekstra and collaborators. In some later experiments, we just mixed them at the appropriate level under controlled conditions, and found that if we added excess lipid or cells to that suspension we didn't get relief of quenching. If R-18 is in a micellar form, it would be quenched, and if it gets into any kind of membrane from that micellar form, it would be dequenched. We didn't see any increase of R-18 fluorescence, when we just mixed the virus up with cell membranes, say, in the cold. One of the reasons we chose that was because of extremely low yields when one passes them through a column, and the dissociation of the gp120. So we chose to take that route. As far as the difference between lymphocyte membranes and erythrocyte ghosts: I'm not sure of the exact lipid compositon, but one difference that one would expect from a ghost membrane is that the lipid asymmetry may have been lost during the preparation of the ghost, so you may have negatively charged lipids on the outer leaflet. That's one possibile difference. And you're right in that we don't know what "receptor" it's binding to on erythrocyte ghosts, versus lymphocytes. We've had a very difficult time showing CD4-specific binding of these labeled viruses, although we've shown that there is gp120 in the virus preparation. Nevertheless, Dr. Sinangil has observed binding to and fusion with $CD4^{+}$ cells (FEBS Lett. 239:88-92). Perhaps it depends on the virus preparation, so we are examining this issue in detail.

Papahadjopoulos (UCSF): Nejat, I'm sure that you have thought of this question, but I wanted to bring it up as a point of discussion. Is it known what the effect of the fluorescent probe is on the infectivity of the virus, and is there any experience with any other viruses to lead us on that? Because that could be a crucial point as to whether the assay is a valid one, with respect to mechanisms of infectivity and entry.

Düzgünes: Chuck Larsen and Krystyna Konopka in my laboratory have performed those experiments and found that labeled virus is still infectious, and binds to liposomes to the same extent as unlabeled virus.

Sinangil (Columbia University): We used R-18 to measure fluorescence dequenching for different kinds of cells using different strains of HIV, and the biological activity is intact after labeling. We followed immuno-fluorescence using the labeled virus in infection experiments, and it turned out to be okay.

Düzgünes: Just on that point, one doesn't necessarily inhibit the fusion activity of a virus by killing the virus. For example, we have worked on ß-propio-lactone-inactivated influenza virus, and it still fuses at low pH. But obviously, it is important to show whether or not the labeled virus is infectious.

Fuller (University of Michigan): We're trying to actually set this up with Herpes simplex and it's been used with Epstein-Barr virus, and it does not affect infectivity. I have two questions. One, how do you determine that you have less than ten percent of the R18 incorporated, and two (I don't think you've probably done this), but will antibodies to either HIV or SIV inhibit fusion in general, and would those antibodies inhibit your fusion here with the liposome? That might help at least a little bit pinpoint biological relevance.

Düzgünes: Under the conditions we use, almost all the probe is incorporated into the viral membrane. We choose the added probe concentration accordingly. We need to be careful about not overloading the membrane.

Larsen (UCSF): We haven't done lipid extractions to determine the exact molar ratio of R18 to viral lipid. It's based on a paper by Aloia et al. (Proc. Natl. Acad. Sci. USA 85:900-904), describing the lipid to protein ratio for HIV, which we're assuming for our strain and for the SIV. We're adding it at what we assume to be approximately six percent of the outer, or three percent of the total lipid. It's three percent of the total lipid that we think, based on the protein concentration, was in the virus.

Düzgünes: We haven't done experiments with antibody. We're ‘hunting’ for antibodies. So, if anybody has an antibody they'd like to share...

Mills: To inject a theological note into this conference: Viruses are dead, because they're not autonomously replicating. But people still talk about killing them, and the most elegant way to do that, if you want to address that issue, is to use one of the psoralen-UV systems, because that's highly selective for viral nucleic acid, and it's been shown in a number of systems that it doesn't disrupt any of the surface structure. So that's what I would probably suggest if you want to do that.

Bassiri (UC Davis): As an alternative for use of antibody, there are, I think, mutants of [the] fusogenic region of SIV. Is it possible to use those mutants to identify whether the model you are proposing is actually correct? In other words, proposing a model for the fusion of your HIV or SIV in the gp41 or 32 region?

Düzgünes: I think that would be an excellent experiment. The only problem is being able to produce enough virus for those studies. We've had a lot of difficulty with producing large quantities of virus; so that would be the rate-limiting step for such a virus preparation. We'd certainly be very interested in doing that.

Sodroski (Harvard): I'll be talking about an infectious clone of HIV-1 that is decreased in fusogenic potential, so theoretically those kinds of studies could be done. The other thing is that the antibodies against post-CD4 binding steps, that inhibit post-CD4 binding steps, are available from the National AIDS Repository, so they're easy to get.

Düzgünes: Can these mutants that you've developed be grown like the parent virus?

Sodroski: Yes.

Düzgünes: We would be very interested in performing those experiments.

# ROLE OF THE HUMAN IMMUNODEFICIENCY VIRUS TYPE I ENVELOPE GLYCOPROTEIN IN CYTOPATHIC EFFECT

Joseph Sodroski, William Haseltine[†] and Mark Kowalski

Laboratory of Human Retrovirology, Department of Pathology
Harvard Medical School, Dana-Farber Cancer Institute and
[†]Department of Cancer Biology, Harvard School of Public Health
Boston, Massachusetts 02115

The pathogenesis of acquired immune deficiency syndrome (AIDS) is characterized by a depletion of CD4-positive lymphocytes (for review, see Fauci, 1988). HIV-1 infection of CD4-positive lymphocytes and, to a lesser extent, monocyte/macrophages in tissue culture is accompanied by cytopathic effects (Popovic et al., 1984). The cytopathic effects follow a temporal course with the initial phase consisting of a period of formation of multinucleated giant cells or syncytia. This is followed by a phase of single cell lysis, which in turn is followed by a period of host cell resistance to cytopathic effect. In the latter phase, cells often continue to produce high titers of viruses that are cytopathic for fresh target cells, without undergoing obvious cytopathic effects.

Early studies into the molecular basis for HIV-1 cytopathic effect indicated that the novel retroviral genes of HIV-1 were not associated with cytopathogenicity independent of their effects on rate of virus replication. Like all retroviruses, HIV-1 possesses long terminal repeats and gag, pol and env genes encoding structural and replicative proteins that are incorporated into virions. HIV-1 possesses six additional genes called vif, vpr, vpu, tat, rev and nef (Sodroski and Haseltine, 1988). Deletions of the vif, vpr, vpu and nef genes yield replication-competent viruses that exhibit both syncytium-forming ability and single cell lytic capacity (Sodroski et al., 1986; Terwilliger et al., 1986; Fisher et al., 1986). The tat and rev genes are essential for HIV-1 replication, but do not induce cytopathic changes in cells susceptible to HIV-1 cytopathic effects (Dayton et al., 1986).

The apparent lack of relationship between the novel genes of HIV-1 and cytopathic effect focused attention on the structural HIV-1 gene products as mediators of target cell killing. In particular, the envelope glycoproteins of HIV-1 have been implicated in both forms of cytopathic effect observed in vitro. The HIV-1 envelope glycoproteins are synthesized from a singly spliced messenger RNA. The 88-kD envelope precursor is glycosylated to a 160-kD form containing high mannose carbohydrate chains (Allen et al., 1985; Robey et al., 1985). This gp160 precursor is subsequently cleaved to a gp120 exterior glycoprotein, which binds to the CD4 virus receptor, and a gp41 transmembrane envelope glycoprotein (Allen et al., 1985; Robey et al., 1985; Veronese et al., 1985).

Expression of the HIV-1 envelope glycoproteins in CD4-positive cells is sufficient to induce the formation of multinucleated giant cells, which develop pyknotic nuclei and eventually die (Sodroski et al., 1986; Lifson et al., 1986). Syncytium formation is not observed when similar levels of HIV-1 envelope glycoproteins are expressed in CD4-negative cells. Syncytium formation in CD4-positive cells is inhibited by monoclonal antibodies to CD4 that block gp120-CD4 interaction, indicating that syncytium formation is dependent upon this interaction. If CD4-negative cells expressing the HIV-1 envelope are cocultivated with CD4-positive lymphocytes, syncytium formation occurs. This observation may be relevant to in vivo pathogenesis, where cells such as monocyte/macrophage derivatives that may serve as a viral reservoir could nucleate the formation of syncytia with CD4-positive lymphocytes. In this manner, even uninfected lymphocytes may be killed. Syncytium formation involving macrophages has been observed in lymph nodes of macaques infected with simian immunodeficiency virus, an immunodeficiency virus closely related to HIV-1 that causes a disease similar to AIDS (Preston Marx and Murray Gardner, personal communication).

Mutagenesis of the HIV-1 envelope glycoprotein has defined regions of the molecule important for the process of syncytium formation (Kowalski et al., 1987). Deletion and in-frame linker insertion mutagenesis was performed on the envelope gene, which was subsequently analyzed for capability to express properly glycosylated and processed envelope precursor (gp160) and mature envelope glycoproteins (gp120 and gp41). The envelope products were also analyzed for their ability to induce the formation of syncytia in CD4-positive lymphocytes. The mutations introduced resulted in several phenotypes:

1. Some mutations in the gp120 - coding section of the envelope gene result in lack of processing of the gp160 envelope precursor to gp120 and gp41 products. These mutants

do not induce the formation of syncytia. The mutant gp160 precursor appears to be properly glycosylated.

2. Mutations in the amino-terminal half of either the gp120 or gp41 proteins result in lack of association of the gp120 and gp41 proteins on the cell surface. This association is not mediated by covalent disulfide bonds. These mutants do not induce the formation of syncytia.

3. Mutations in any one of three regions in the carboxy-terminal half of gp120 that are well-conserved among HIV-1, HIV-2 and SIV result in a loss of the ability of gp120 to bind to the CD4 virus receptor. These regions are separated by hypervariable regions that can be mutated without affecting CD4 binding. This suggests that regions of gp120 important for CD4 binding are discontinous and require proper tertiary structure for alignment. Mutations that disrupt gp120-CD4 binding also destroy syncytium-forming ability.

4. Mutations in the amino terminus of gp41 result in envelope glycoproteins that are expressed on the cell surface and that bind CD4 receptor but that do not form syncytia. Apparently, a function critical for the fusion of membranes is localized in the hydrophobic amino-terminal region of gp41.

5. Large deletions or substitutions in the carboxyl terminus of gp41, a region that is unusual for retroviral transmembrane proteins, do not affect syncytium forming ability or cytopathicity of the virus. However, viruses with these changes in gp41 exhibit delayed replication compared with wild-type viruses.

6. Deletion of the hydrophobic membrane-spanning region of gp41 results in envelope products that exhibit inefficient cleavage at the gp120-gp41 junction, that are preferentially found in the supernatant of envelope-expressing cells, and that do not induce the formation of syncytia, even though CD4-binding is not detectably altered.

The mutagenic studies suggest a functional model of the HIV-1 envelope glycoproteins, the key features of which are:

1. The gp41 glycoprotein is anchored in the envelope-expressing cell membrane via a hydrophobic region near the middle of the molecule.

2. The gp41 and gp120 molecules are non-covalently associated via interactions between the amino-terminal halves of these glycoproteins.

3. Binding to CD4 occurs via conserved, discontinuous regions located near the carboxyl terminus of gp120.

4. Events that lead to the fusion of cell membranes occur. At least some of these events are mediated by a hydrophobic region at the very amino terminus of gp41. This region, by analogy with similar regions in para- and ortho-myxovirus fusion proteins, may directly interact with the target cell membrane.

HIV-1 envelopes containing mutations in the gp41 amino terminus that affect the ability to fuse membranes have been used to investigate the mechanisms of single cell lysis by HIV-1. Two mutants, 517A and 517B, were generated, containing four or six amino acids, respectively, inserted into the gp41 amino terminus. The 517B insert contains a positively charged residue and thus, is more disruptive of the hydrophobicity of the gp41 amino terminus than is the 517A mutation. These mutations do not appreciably affect the synthesis, processing, cell surface expression, stability, subunit association or CD4-binding ability of the mutant envelopes compared with the wild-type HIV-1 envelope. Both mutants exhibit decreased ability to fuse membranes in a syncytium formation assay. The 517B mutant is completely devoid of detectable syncytium-forming activity, whereas the 517A mutant exhibits between 1-5% of wild-type syncytium-forming ability (Kowalski, 1987).

The effect of these mutations on replication and in vitro cytopathogenicity of HIV-1 was assessed by introducing these mutant env genes into the infectious HIV-1 provirus contained on plasmid pHXBc2. In this manner, isogenic HIV-1 proviruses save for the mutations in the envelope fusion region were created. Upon transfection of the human T lymphocyte line Jurkat with the parental and mutant pHXBc2 plasmids, virus replication and cytopathic effect were assessed. Virus replication was assayed by measuring the amount of viral proteins synthesized per viable cell in the culture (radioimmunoprecipitation), reverse transcriptase and p24 gag protein levels in the cell supernatants, and HIV-specific cell surface fluorescence. Viral cytopathic effect was assayed by counting the number of syncytia and by scoring the percentage of single cells that excluded viral dyes, in addition to measuring total viable cell counts.

The replication ability of the 517B mutant in Jurkat cells is markedly attenuated compared with the wild-type virus. Upon transfection of COS-1 cells, the production of viral proteins and virion particles is indistinguishable between the 517B mutant and the wild-type virus. The binding of the 517B mutant virions produced in COS-1 cells to the CD4 receptor is also comparable to that of the wild-type virus. These results suggest that a mutation that markedly attenuates the membrane-fusing capacity of the HIV-1 envelope affects virus replication, probably at an early step in virus replication.

In contrast to the attenuated replication exhibited by the 517B virus, the 517A mutant virus replicated in transfected Jurkat cells at a rate comparable to that of the wild-type virus, using the parameters detailed above to measure virus replication. The cytopathic effects of the 517A virus on the Jurkat cells were markedly diminished compared to those of the wild-type virus. Jurkat cultures transfected with the wild-type proviral DNA demonstrated a progressive decline in total viable cell counts. By contrast, the Jurkat culture infected with the 517A virus continued to grow exponentially, almost as rapidly as the mock-transfected control cells. Measurements of syncytium formation and single cell killing indicated that the 517A virus was attenuated both in the ability to induce the formation of syncytia and in the ability to lyse single cells. Thus, all of the in vitro cytopathic effects accompanying HIV-1 infection can be attenuated by a mutation that decreases, but does not abolish, membrane-fusing ability of the viral envelope.

These results indicate that the membrane-fusing ability of the HIV envelope plays a role in virus replication in T lymphocytes and in the direct cytopathic consequences of HIV-1 infection. The observation that some changes affecting envelope fusogenicity (e.g., the 517A mutation) affect cytopathogenicity without markedly affecting virus replication suggests that more envelope-receptor interactions are likely to be involved in cytopathic membrane fusion events than in virus entry and replication. Thus, changes in the efficiency of envelope-mediated membrane fusion would exert more dramatic effects on processes requiring many independent envelope-receptor interactions than on processes requiring only a few such interactions.

Since the levels of viral protein expression for both the wild-type and 517A viruses were identical at all time points examined and never exceeded 5% of the total host cell protein synthesis, these results also indicate that the model for HIV-1 single cell killing proposed by Somasundarin and Robinson (1988) is not general. These investigators suggest that viral protein overexpression and pre-emption of host cell protein synthesis are responsible for single cell lysis. Our results suggest that high levels of viral protein expression and virion release are not sufficient for cytopathic effect, in agreement with DeRossi et al. (1986) and Fisher et al. (1986).

In summary, the ability of the HIV-1 envelope to fuse membranes has been implicated in both forms of in vitro cytopathic effect, i.e., syncytium formation and single cell lysis. The relevant compartment for envelope-receptor interactions involved in single cell lysis (intracellular, cell surface, or virion-cell surface) remains to be established. The contribution of virus-mediated cell killing to in vivo pathogenesis of AIDS is a second critical area for further investigation.

## Acknowledgements

We thank A. Dayton, M. Wojtkowski, E. Terwilliger, T. Dorfman and L. Basiripour for contributing to some of the experiments described herein. We thank F. Wong-Staal, R.C. Gallo, T.H. Lee, M.F. McLane, and M. Essex for reagents, and T. Parks for manuscript preparation. J.G.S. is supported by the Leukemia Society of America, the Department of the Army and the National Institutes of Health. W.A.H. is supported by the State of Massachusetts and the National Institutes of Health. M.K. is supported by the American Foundation for AIDS Research.

## References

J. Allen, T.H. Lee, M.F. McLane, J.G. Sodroski, W.A. Haseltine, and M. Essex, Identification of the major envelope glycoprotein product of HTLV-III, Science 228:1091 (1985).

A.I. Dayton, J.G. Sodroski, C.A. Rosen, W.C. Goh, and W.A. Haseltine, The trans-activator gene of the human T-cell lymphocyte virus type III is required for replication, Cell 44:941 (1986).

A. DeRossi, G. Franchini, A. Aldovini, A. DelMistro, L. Chieco-Bianchi, R.C. Gallo, and F. Wong-Staal, Differential response to the cytopathic effects of HTLV-III superinfection in T4+ (helper) and T8+ (suppressor) T-cell clones transformed by HTLV-I, Proc. Natn. Acad. Sci. U.S.A. 83:4297 (1986).

A. Fauci, The Human Immunodeficiency Virus: Infectivity and Mechanisms of Pathogenesis, in: "AIDS 1988," Ruth Kulstad, ed., American Association for the Advancement of Science, Washington, D.C. (1988).

A. Fisher, L. Ratner, M. Mitsuya, L. Marselle, M.E. Harper, S. Broder, R.C. Gallo and F. Wong-Staal, Infectious mutants of HTLV-III with changes in the 3' region and markedly reduced cytopathic effects, Science 233:655 (1986).

M. Kowalski, J. Potz, L. Basiripour, T. Dorfman, W.C. Goh, E. Terwilliger, A. Dayton, C. Rosen, W.A. Haseltine, and J.G. Sodroski, Functional regions of the human immunodeficiency virus envelope glycoprotein, Science 237:1351 (1987).

J. Lifson, M. Feinberg, G. Reyes, L. Rabin, B. Banapour, S. Chakrabarti, B. Moss, F. Wong-Staal, K. Steimer, and E. Engleman, Induction of CD4-dependent cell fusion by the HTLV-III/LAV envelope glycoprotein, Nature 323:725 (1986).

M. Popovic, M.G. Sarngadharan, E. Read, and R. C. Gallo, Detection, isolation, and continous production of cytopathic retroviruses (HTLV-III) from patients with AIDS and pre-AIDS, Science 224:497 (1984).

W.G. Robey, B. Safai, S. Oroszlan, L. Arthur, M. Gonda, R.C. Gallo and P. Fischinger, Characterization of envelope and core structural gene products of HTLV-III with sera from AIDS patients, Science 228:593 (1985).

J. Sodroski, W.C. Goh, C.A. Rosen, A. Tartar, D. Portetelle, A. Burny, and W.A. Haseltine, Replicative and cytopathic potential of human T-lymphotropic virus type III with sor gene deletions, Science 240:1549 (1986).

J. Sodroski, W.C. Goh, C.A. Rosen, K. Campbell, and W.A. Haseltine, Role of the HTLV-III envelope in syncytium formation and cytopathicity, Nature 322:470 (1986).

J. Sodroski and W.A. Haseltine, The Molecular Biology of HIV Replication and Pathogenesis, in: "AIDS 1988", Ruth Kulstad, ed., American Association for the Advancement of Science, Washington, D.C. (1988).

E. Terwilliger, J.G. Sodroski, C.A. Rosen, and W.A. Haseltine, Effects of mutations within the 3'-ORF region of the HTLV-III/LAV on replication and cytopathicity, J. Virol. 60:754 (1986).

F.D. Veronese, A.L. DeVico, T. Copeland, S. Oroszlan, R.C. Gallo and M.G. Sarngadharan, Characterization of gp41 as the transmembrane protein coded by the HTLV-III/LAV envelope gene, Science 229:1402 (1985).

DISCUSSION

Milich (UC Davis): Do you have direct evidence that giant cells in SIV-infected macaque lymph nodes express SIV antigens?

Sodroski: Murray Gardner at UC Davis and others at the New England Primate Center have demonstrated that these cells are SIV-antigen-positive, and are not associated with opportunistic infections by, for example, cytomegalovirus.

Banapour (UC Davis): Since the 517B virus cannot perform a second cycle of infection, does the virus disappear following transient expression?

Sodroski: We observe a very low level of persistent expression of 517B viral proteins, which I attribute to a low level of replication. When large deletions in env or pol are made, those replication-incompetent viruses express proteins transiently after transfection and then disappear completely. So it's likely that the 517B mutation allows some level of residual replication.

Banapour: And you attribute that just to DNA staying in the...

Sodroski: I attribute that to a low level of replication. I think that if we make what must be a replication incompetent virus by deleting a large portion of the envelope, if you look at those pro-viruses and transfect those into cells, you see transient expression, and then it goes away completely. It's totally undetectable. So the fact that we see a very low level of persistent protein with the 517B mutation suggests that there's probably a low level of replication. It's not enough to exceed the replication cycle of the cell, but it's enough to keep up with it. So it's very low, basically it goes almost undetected by densitometry, but if you look at it by eye, you can just see a little bit of protein being made. So I think there's some residual level of replication left.

Pauza (Salk Institute): Do cells that express gp120 fuse with other cells that express gp120?

Sodroski: Cells that express gp120, or are infected, tend to have their CD4 receptor blocked by interference and therefore, are not good targets for syncytium formation. Later in the infection, the absolute amount of CD4 in the culture is decreased, so more than simple interference is occurring. These cells do not express CD4, so they will not act as fusion partners.

Pauza: Right, because that's a sort of standard model for how cells survive infections. They achieve interference immunity faster than they undergo the cytopathic effect. That's clear.

Sodroski: And, what you see though is that even when there's not enough CD4 for syncytium formation, there must be enough for single-cell lysis because that can occur at days 4 through 9 when syncytium formation is already over.

Pauza: Can you observe single-cell lysis in the presence of soluble CD4?

Sodroski: Those experiments are in progress, and I do not have an answer yet.

Pauza: We've done some experiments where we tried to look at the relationship between multiplicity of infection and single-cell lysis, and unfortunately, after listening to this talk, I realized one of the problems probably is the amount of soluble gp120 that's in those high-titer inocula. But in general, we thought it was sort of related to multiplicity. Have you observed a relationship between multiplicity of infection and single-cell lysis?

Sodroski: The size and number of syncytia following HIV-1 infection are maximal at a certain multiplicity of infection. At lower multiplicities of infection, less syncytia are formed, probably due to fewer numbers of envelope-expressing cells. At higher multiplicities of infection, single cell lysis is more prominent than syncytium formation, which is decreased secondary to interference by the large amount of envelope glycoproteins in the culture.

# THE ASSEMBLY OF THE HIV-1 ENV GLYCOPROTEIN INTO DIMERS AND TETRAMERS

Robert W. Doms*^, Patricia L. Earl*, and Bernard Moss*

*Laboratory of Viral Diseases, National Institute of Allergy and Infectious Diseases, and ^Laboratory of Pathology, National Cancer Institute, National Institutes of Health, Bethesda, MD 20892

## ABSTRACT

The envelope (env) glycoprotein of human immunodeficiency virus 1 (HIV-1), initially synthesized as a precursor molecule termed gp160, is cleaved into two noncovalently associated subunits prior to delivery to the plasma membrane. We have studied the oligomeric structure of this protein using chemical cross-linking, velocity gradient sedimentation, and SDS-resistance. We find that gp160 forms stable homodimers after synthesis. After cleavage to gp120/gp41 the molecule becomes less stable to detergent solubilization and centrifugation but remains dimeric. Interactions between the 129 amino terminal residues in the ectodomains of adjoining gp41 subunits are both sufficient and necessary for assembly. In addition, tetramers composed of two dimers were also formed. Larger structures were not observed. The tetrameric paramyxovirus F protein, which has structural and functional similarities to the HIV-1 env protein, also forms a dimer of dimers.

## INTRODUCTION

The envelope (env) glycoprotein of HIV-1 is responsible for binding virus to the surface of cells that express CD4 and mediating fusion between the viral and plasma membranes (1-7). Elucidating the structure of gp160 is important for understanding both of these functions. The env protein is initially synthesized as a single polypeptide precursor, gp160 (8,9). It is core-glycosylated and folds into a conformation competent for CD4 binding shortly after synthesis (10). Prior to delivery to the plasma membrane, gp160 undergoes a single post-translational proteolytic cleavage which is essential for infectivity. The cleavage creates two noncovalently associated subunits; gp120 which is external to the membrane and gp41 which spans the viral membrane (8,9,11). The newly generated amino terminal domain of gp41 is highly conserved and hydrophobic and is likely to play a role in membrane fusion (12). In this regards, the env protein is similar to many other viral envelope proteins whose membrane fusion activities are contingent upon an analogous cleavage which likewise generate fusogenic domains (13,14).

While the roles of the HIV-1 env protein in the virus life cycle are becoming better understood, little is known about the molecule's higher order structure. A feature common to many viral and cellular membrane proteins is that they are oligomeric (15,16). Assembly into higher order forms has important consequences. For example, correct folding and assembly is required for transport from the ER (15,16). If a protein fails to fold and/or assemble correctly due to a mutation, adverse cellular conditions, or simply by chance, the molecule is retained in the ER and eventually degraded (17). In addition, a protein's function is intimately related to its tertiary and quaternary structure. Thus, to correlate the functions of the HIV-1 env protein with its sturcture, we have studied its quaternary structure by a variety of techniques. We find that gp160 forms stable homodimers after synthesis, and that at least a fraction of these molecules assemble into a higher order structure, most likely a tetramer composed of two dimers.

## MATERIALS AND METHODS

### Cells and viruses

The HIV-1 env gene from the BH8 isolate was inserted into vaccinia virus as described (18,19). The following viruses were used: vPE16 which encodes the full length, wild type molecule; vPE12 which lacks the cleavage site between gp120 and gp41 (20), vPE17 which lacks the 104 carboxy terminal amino acids of the molecule (constituting most of the cytoplasmic domain), and vPE18 which contains all of gp120 but only the first 129 residues of gp41 (20). BSC1 cells were grown in MEM with 2.5% serum. Confluent cells were infected with 20-30 plaque forming units of recombinant vaccinia virus per cell. Pulse-chase experiments were performed approximately 5 hours post-infection.

### Gradient centrifugation

Cells were lysed in 250mM octyl glucoside or 1.0% (w/v) Triton X-100. After 10min on ice, the lysates were spun for 10min at 10,000g to remove cellular debris and virus particles. The clarified lysates were then loaded onto continuous 5-20% (w/v) sucrose gradients made in 100mM Tris HCl pH 8.0, 100mM NaCl, 1mM $CaCl_2$ and 25mM octyl glucoside or 0.1% (w/v) TX100. Centrifugations were performed at 4°C, typically for 20h at 40,000 rpm in the SW40 rotor or for 14 hours at 42,000 rpm in the SW50.1 rotor unless otherwise indicated. After centrifugation, fractions were collected from the bottom of each tube using a Beckman Fraction Recovery System. Generally, 24 fractions were collected from SW40 gradients and 15 from SW50.1 gradients. Sedimentation coefficients were calculated as described by McEwen (21) as well as by the method of Martin and Ames (22), using the influenza hemagglutinin (23) and vesicular stomatitis virus G protein (24) as standards. The distribution of env protein across the gradient was analyzed by either immunoprecipitating the env protein with a polyclonal antibody to gp120 or by acid precipitation followed by SDS-PAGE and western blotting.

Chemical cross-linking

Cross-linking reagents were obtained from Pierce Chemical Co. Cross-linking on the cell surface was performed by rinsing the cells with phosphate buffered saline, pH 8.0 at room temperature. The cross-linker, dissolved in DMSO at a 100-fold concentration, was then added and incubated with the cells for 15min. The cells were then washed with PBS and lysed with TX100 and 10mM glycine. E. coli ß-galactosidase (Calbiochem) at a concentration of 10mg/ml in PBS was cross-linked with 1mM EGS for 15min at room temperature and used as a MW standard.

## RESULTS

To examine the quaternary structure of the HIV-1 env protein, we used recombinant vaccinia viruses which expressed the wt or mutant forms of the molecule. In general, proteins expressed by recombinant vaccinia viruses are processed and transported to the cell surface normally. This holds true for the HIV-1 env protein as well. The gp160 synthesized in cells infected with a recombinant virus expressing the wt molecule (vPE16) is glycosylated, cleaved into gp120 and gp41 subunits, expressed on the cell surface with high efficiency, and mediates fusion with CD4 bearing cells - thus allowing syncytia formation (19,25). In this study we also examined the quaternary structures of several mutant forms of the env protein, including a molecule which is not processed to gp120/41 (vPE12) and, as a result, cannot induce syncytia (20). Two mutants with truncations in gp41 were also studied including vPE17 which lacks the cytoplasmic domain, and vPE18 which has only the first 129 amino acids of the gp41 ectodomain (20). Because of its deletion, vPE18 is secreted from cells (20).

It is often difficult to establish the oligomeric structure of membrane proteins without ambiguity. Each of the most commonly used biochemcial techniques has its own drawbacks. As a result, it is important to compare the results obtained with one technique to those obtained by another. We initially employed sucrose velocity gradient sedimentation to examine the quaternary structure of the env protein. Cells infected the recombinant vaccinia viruses were lysed in octyl glucoside or Triton X-100 and the lysates centrifuged on 5-20% (w/v) sucrose gradients. Following fractionation, the distribution of the env glycoprotein across the gradient was determined by SDS-PAGE and western-blotting with an antibody to gp120 (Fig. 1). In cells expressing the wild type molecule (vPE16), most of the gp160 was found in fractions near the bottom of the tube with a sedimentation coefficient of approximately 10.8S. While most of the gp160 in these fractions migrated as monomer in SDS-PAGE, some of the protein migrated with a much higher apparent molecular weight. When analyzed on low percentage acrylamide gels, the MW of this complex was estimated to be approximately 320kd. This, and its reactivity with a monoclonal antibody to gp120 suggested that this band represented an SDS-resistant dimer of gp160 molecules. A similar degree of SDS-resistance has been observed with the env proteins of HIV-2 and SIV (27) as well as several other multimeric membrane proteins (23,28). Since SDS-resistance is unlikely to arise from detergent induced aggregation, it is likely to reflect a specific, stable interaction (26).

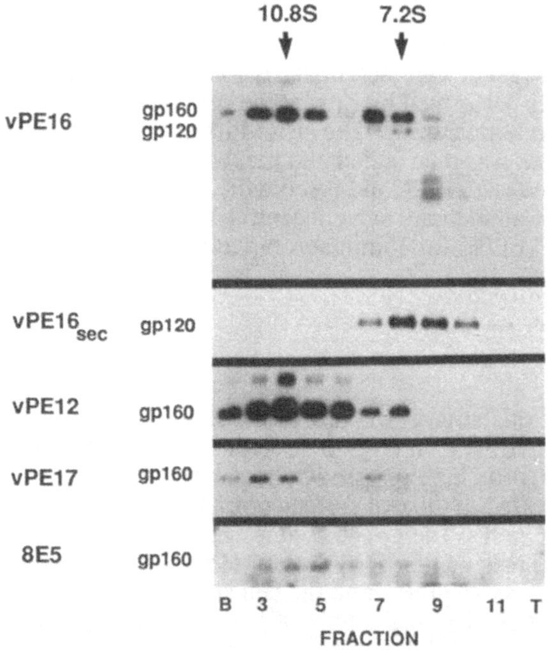


Fig. 1. Velocity gradient sedimentation of HIV-1 env glycoproteins. BSC1 cells were infected with recombinant vaccinia viruses encoding the wild-type (vPE16), processing mutant (vPE12), or cytoplasmic domain mutant (vPE17). Twenty hours after infection, the cells were lysed as described in Materials and Methods and the lysates placed atop a continuous sucrose gradient. After centrifugation, the distribution of env protein across the gradient was determined by acid precipitation, SDS-PAGE, and western blotting with a monoclonal antibody to gp120 (45).

A smaller amount of gp160 was recovered from a 7.2S peak. Protein from this peak gave a single monomeric gp160 band in SDS-PAGE, demonstrating that the dimer observed in the 10.8S fractions did not arise from disulfide exchange during sample preparation. In addition, gp120 and gp41 were also recovered from the 7.2S peak. Secreted gp120 applied to a separate gradient also sedimented at 7.2S. Finally, when cell lysates were boiled in the presence of SDS and reducing agent prior to centrifugation, greater than 90% of the gp160 was recovered from the 7.2S peak. Lighter sedimenting forms were not observed. Thus, we concluded that the 7.2S material was monomeric.

The distribution of gp160 from cells expressing the unprocessed molecule (vPE12) as well as the cytoplasmic tail mutant was similar to that seen with the wt molecule. Most of the protein sedimented at 10.8S while a smaller amount was recovered from the 7.2S peak. Once again, nondissociated gp160 dimer from the 10.8S fractions was observed by SDS-PAGE. By using the influenza hemagglutinin (9S) and vesicular stomatitis virus G protein trimers (8S) and monomers (4S) as markers, we estimated the molecular weights of the 10.8S and 7.2S proteins to be

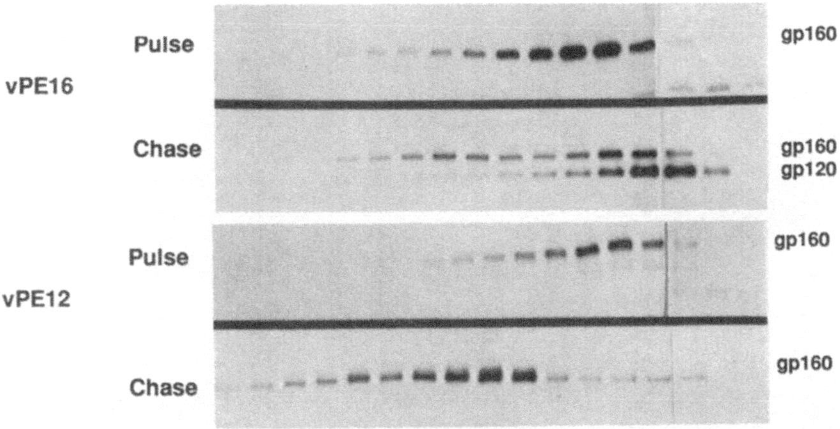


Fig. 2. Post-translational assembly of env protein dimers. Cells expressing the wt (vPE16) or cleavage-minus (vPE12) molecules were pulse labeled for 20 min and then lysed immediately or after a 4 hour chase. The lysates were centrifuged for 23 hrs at 40,000 rpm in a SW40 rotor at 4°C. Twenty four fractions were collected from each gradient, and the env protein in each was immunoprecipitated with rabbit antisera (PB33, a gift from P. Berman, Genentech) and analyzed by SDS-PAGE.

approximately 310kd and 167kd, respectively. This, coupled with the presence of SDS-resistant gp160 dimers suggested that protein from the 10.8S peak was dimeric. Similar results were obtained if Triton X-100 or CHAPS were used in place of octyl glucoside.

We next employed a pulse-chase protocol to determine if sedimentation of gp160 in the 10.8S peak was due to post-translational assembly. Most proteins assemble post-translationally within an hour after synthesis (15,16). Cells infected with vPE16 were pulse labeled for 20min with 35S-methionine and lysed immediately or after a 4 hour chase. After centrifugation, fractions were immune precipitated with rabbit antisera to gp120. As shown in Figure 2, most of the gp160 sedimented in the 7.2S peak immediately following a 20min pulse-label. After a 4 hour chase, much of the gp160 sedimented in the 10.8S region. However, the gp120 generated during the chase period was recovered almost exclusively in the monomeric 7.2S region. This experiment showed that gp160, synthesized as a monomer, formed dimers post translationally. After proteolytic cleavage to gp120/gp41, the dimers either dissociated to monomers or became less stable to detergent solubilization and centrifugation. It has been suggested that the HIV-2 and SIV env proteins form dimers during their biosynthesis but revert to and remain in monomeric form in the trans Golgi (27). Other proteins, such as some strains of influenza hemagglutinin, oligomerize shortly after synthesis but do not become stable to detergent solubilization and centrifugation until much later (36).

To see if the env protein remained oligomeric after activation to gp120/gp41, we performed chemical cross-linking on intact cells infected with vPE16 and

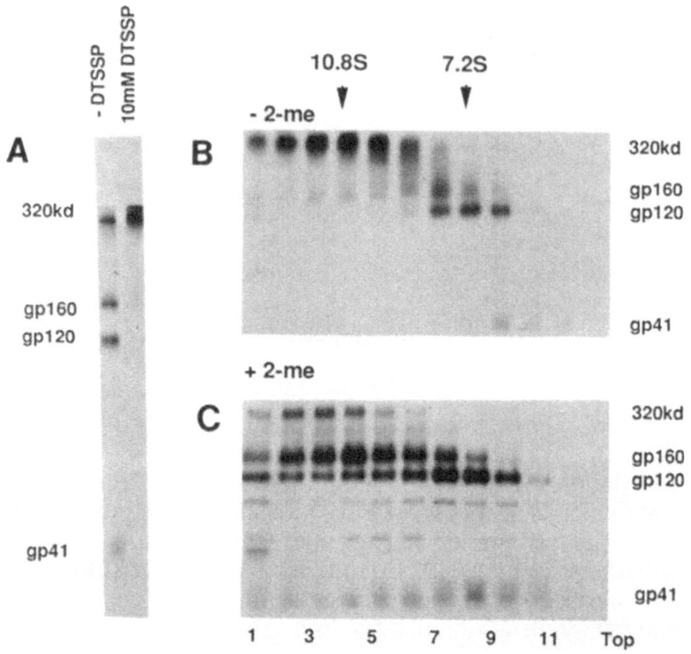


Fig. 3. Cross-linking of cell surface env protein. Cells expressing the wt molecule were labelled with methionine for 16 hours. The cells were then washed with PBS, pH 8.0, after which 10mM DTSSP was added. Cross-linking was allowed to proceed for 30 min on ice. The cells were then lysed and the env proteins immunoprecipitated with HIV positive human sera (Panel A). An aliquot of the lysate was loaded onto a 5-20% (w/v) sucrose gradient and centrifuged as in Figure 1. The distribution of the env protein across the gradient was determined by SDS-PAGE under both non-reducing (B) and reducing (C) conditions.

metabolically labeled with 35S-methionine. Cross-linking on the cell surface stabilizes multimeric proteins prior to detergent solubilization which might otherwise cause dissociation. Since there is no way in advance to determine which reagent will most efficiently cross-link any given protein, a battery of both cleavable and non-cleavable cross-linking reagents was employed. As shown in Fig. 3A, the cleavable cross-linking reagent 3,3'-dithiobis(sulfosuccinimidyl) propionate (DTSSP) efficiently cross-linked gp160, gp120 and gp41 into a single high molecular weight band of approximately 320kd. When this band was excised from the gel and the sample re-run under reducing conditions so as to break the chemical cross-linker, gp160, gp120 and gp41 were the only proteins recovered.

To more fully analyze the 320kd complex, a part of the cross-linked cell lysate was subjected to gradient centrifugation. Gradient fractions were analyzed by SDS-PAGE in the presence or absence of reducing agent. As shown in Figure 3B, the cross-linked material was almost quantitatively recovered from the 10.8S peak. When the same samples were reduced prior to electrophoresis, we found that the cross-linked material consisted of gp120/gp41 as well as gp160 (Fig. 3C). Thus, the env protein exists as a dimer on the cell surface even after activation to the gp120/41 complex.

The results in Figure 2 demonstrate an additional point. While the gp160 remaining after the chase period sedimented mostly in the 10.8S region, some of the gp160 sedimented more quickly, with $S_{20,w}$ values ranging from 14S to 16S. This was more evident in cells expressing the mutant env protein which was not activated to gp120 and gp41 (vPE12). By 4hrs of chase, the protein sedimented at 10.8S but the peak was extremely broad, with a long "leading edge" of material which sedimented as far down as approximately 18S (Figure 2). That some gp160 sedimented more quickly than 10.8S can be explained in several ways. First, some of the protein might aggregate with itself or with cellular proteins. Alternatively, a fraction of the gp160 could assemble into higher order oligomers. Assembly into higher order forms could be inefficient, accounting for the small amount of material sedimenting at greater than 10.8S, or larger structures could be less stable to centrifugation and/or solubilization.

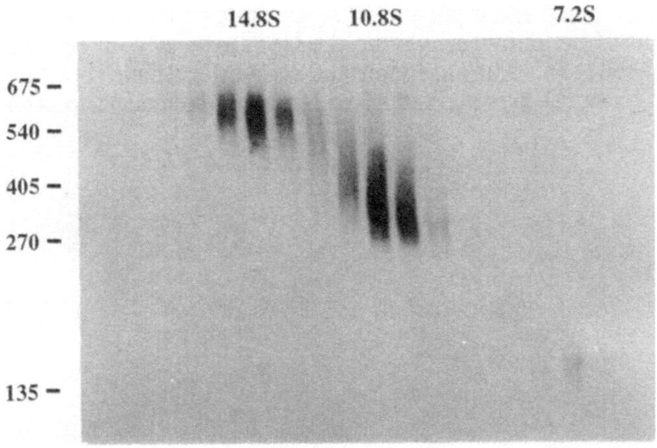


Fig. 4. Cross-linking of the env protein after centrifugation. Twenty hours after infection, cells expressing the wt env protein were lysed and subjected to ultracentrifugation at 40,000 rpm at 4°C for 20 hrs in a SW40 rotor. Following fractionation, each fraction was cross-linked by addition of 5mM EGS for 15 min at room temperature. The reactions were quenched by addition of glycine and the proteins analyzed as in Figure 1 but on a 4% acrylamide gel. E. coli ß-galactosidase, cross-linked with EGS, was used as a MW standard.

To address the structure of the gp160 which sedimented at greater than 10.8S, we employed chemical cross-linking. Cells infected with vPE16 were lysed and the solubilized protein subjected to velocity gradient sedimentation. Afterwards, each fraction was cross-linked with 5mM EGS for 15 min at room temperature. The proteins were then recovered by acid precipitation and analyzed by SDS-PAGE and western blotting. As seen in figure 4, three forms of the protein were recovered. The gp160 in the 7.2S peak could not be cross-linked into higher order structures, confirming its identity as monomer. The gp160 recovered in the 10.8S region was quantitatively cross-linked into a single, broad band with a MW of approximately 320kD. Estimation of the MW of this complex was made by using 4% acrylamide gels and with cross-linked ß-galactosidase as a marker. Thus, the 10.8S material represented dimeric gp160. Finally, gp160 from 14S to 16S was cross-linked into a single band with a MW of approximately 600kD. Rarely, a less intense band was observed which was intermediate between the 10.8S and 14-16S material both in sedimentation coefficient and MW.

The identity of the 600kD band is somewhat uncertain at this time. Both its MW and sedimentation coefficient are between what would be expected for trimeric and tetrameric gp160. We favor the idea that this protein is tetrameric for several reasons. First, dimeric gp160 seems to represent a stable "building block". As noted above and also described in experiments to follow, dimeric gp160 is exceptionally stable and easily isolated. Given this, then it seems far more likely that the 600kd material represents a tetramer of gp160 composed of two dimers rather than a trimer consisting of a dimer and a monomer associated with one another. A "dimer of dimers" apparently represents the structure of both the F and HN envelope proteins of Sendai virus which, like HIV-1, fuses with cellular membranes at neutral pH (29,30). In addition, Pinter and co-workers have reported evidence which suggests that gp41 exists in tetrameric form in isolated virions (31).

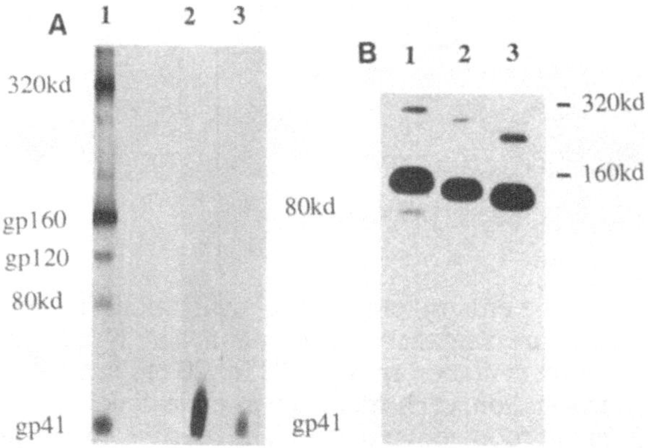


Fig. 5. Cross-linking intermediates. A. Cross-linking was performed as described in Figure 4 but with 1mM DTSSP. Proteins were analyzed on a 5% - 10% gradient gel. Bands migrating at 40kd and 80kd were excised and re-run under reducing conditions so as to break the chemical cross-linker. B. Extracts of cells expressing vPE16 (lane 1), vPE17 (lane 2), or vPE18 (lane 3) were analyzed by SDS-PAGE and western blotting to examine the SDS-resistance of each protein.

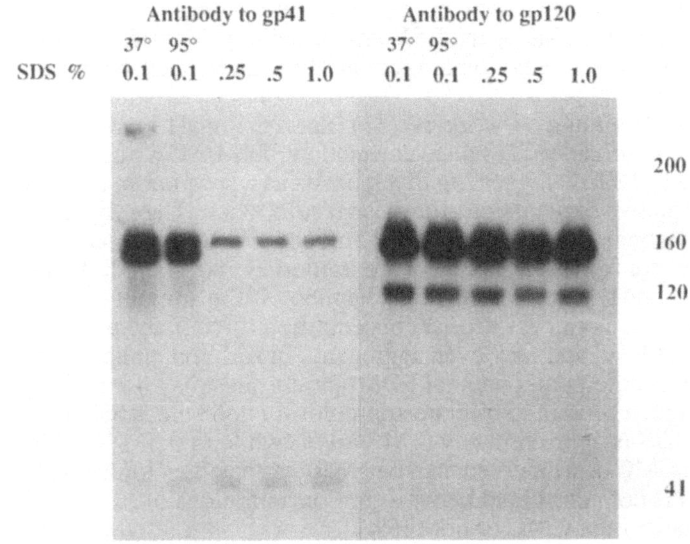


Fig. 6. Effects of low SDS on gp41 migration. Cells expressing wt env protein were lysed with the indicated concentration of SDS. Samples were heated for 5 min at either 95°C or 37°C as indicated and subjected to SDS-PAGE with 0.1% SDS in the running and gel buffers. Identical samples were western blotted with antibody to gp41 or with antibody to gp120.

We next determined which subunit of the env glycoprotein was responsible for dimer stability. Cells infected with vPE16 and metabolically labeled with 35S-methionine were subjected to chemical cross-linking under conditions designed to reveal cross-linked intermediates between the gp41 and gp120 subunit monomers and the final 320kd complex. As shown in Figure 5, a prominent band of approximately 80kd was observed after incomplete cross-linking. When this band was excised and subjected to a second round of SDS-PAGE under reducing conditions, gp41 was recovered. By contrast, we have not yet observed cross-linked gp120 dimers.

As shown in Figures 1 and 2, gp160 displayed partial resistance to SDS-induced dissociation. Resistance to dissociaiton by SDS is a property shared by several other oligomeric membrane proteins and, when present, likely reflects a native interaction (26). Taking advantage of this property, we determined if other forms of the env protein displayed resistance to SDS. We found that vPE17, which lacks the cytoplasmic domain of gp41, gave SDS-resistant dimers as did vPE18 which contains only the amino terminal 129 amino acids of gp41 (Fig. 5). In addition, both of these molecules formed oligomers as determined by velocity gradient sedimentation (Figure 1 and unpublished observations). Thus, neither the transmembrane or cytoplasmic domains of gp41 are absolutely required for dimer formation and stability. However, gp120 recovered from the medium of cells expressing the wt molecule sedimented in monomeric form and gave only monomers in SDS-PAGE (Figures 1 and 2). Thus, we concluded that the amino terminal domain of gp41 is both sufficient and necessary for assembly. In this regard the HIV-1 env protein is similar to the RSV env protein. While RSV is trimeric, it is likewise stabilized by the ectodomain region of its membrane spanning subunit (32).

Recently, Pinter and co-workers (31) found that gp41 extracted from virions with low concentrations of SDS and subjected to SDS-PAGE migrates with a MW of approximately 160kd, suggesting that it exists as a tetramer in the viral membrane. However, alternative explanations for these results exist. The migration of proteins in SDS gels is largely dependent upon the amount of detergent bound to the protein (46,47). If low concentrations of SDS are used to extract a small, hydrophobic protein such as gp41, several things may happen. If the amount of SDS used is below the detergent's critical micellar concentration (CMC), the protein may not be effectively solubilized and may even aggregate. In addition, under limiting concentrations of detergent, potential SDS-binding sites may not be fully saturated, thereby causing the protein to migrate anomolously upon electrophoresis. In water, the CMC for SDS is approximately 0.24% (w/v) while in 0.5M NaCl it is 0.015%. Given that the CMC of a detergent can be greatly affected by ionic conditions and temperature, it is not possible to know if the concentrations of SDS used by Pinter et al. (31) were below the CMC or not (46,47).

We decided to repeat and extend the studies of Pinter et al. (31) by examining the migration of gp160, gp120 and gp41 under low SDS conditions. Cells expressing the wt env protein were extracted with different concentrations of SDS, subjected to SDS-PAGE under low SDS conditions, and western blotted with antibodies to either gp41 or gp120. We also incubated samples at 37°C as well as 95°C. As seen in figure 6, the antibody to gp41 recognized both gp160 and gp41 under normal (1%) SDS conditions. However, at concentrations below 0.25% SDS the amount of monomeric gp41 decreased dramatically and a new band with a MW of approximately 150kd appeared. The intensity of this band was far greater than the original gp41 band even though the amount of protein loaded onto each lane was identical. The most obvious interpretation is that this band represents gp41 and that it is recognized more efficiently by the antibody either because it remains oligomeric or is not extensively denatured. When identical samples were hybridized with an antibody to gp120, it could easily be seen that the migration of gp120 and gp160 was not affected by incubation with low SDS and that the 150kd band consisted solely of gp41. Thus, we were able to confirm the results of Pinter et al.(31), but were unable to conclude whether the 150kd band was the result of aggregation of gp41 in low SDS, anomolous migration, or because the protein exists in a tetrameric state.

To further investigate this interesting phenomenon, we asked what would happen to monomeric gp41 under identical conditions. To prepare monomeric gp41, cell lysates were boiled with reducing agent and 1.0% SDS and then centrifuged on sucrose gradients designed to separate monomeric gp41 from monomeric gp120 and gp160. As seen in Figure 7, gp41 treated in this fashion displayed a sedimentation coefficient of 3S. The sedimentation of marker proteins in a parallel gradient is shown as well. Aliquots from the peak fraction containing monomeric gp41 were then run in SDS-PAGE at different SDS concentrations (Fig. 8). As expected, gp41 subjected to electrophoresis in the presence of 1.0% SDS ran solely in monomeric form. At 0.1% SDS, however, some of the protein migrated with an apparant MW

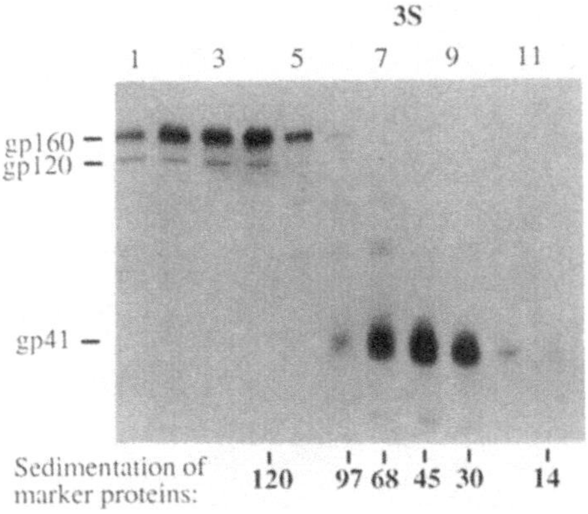


Fig. 7. Generation of monomeric gp41. A cell lysate was boiled for 5 min with 1% SDS and 1% ß-mercaptoethanol and then centrifuged for 20 hrs at 47,000 rpm in the SW55 rotor. After fractionation an aliquot of each fraction was TCA precipitated, subjected to SDS-PAGE, and hybridized with an antibody to gp41. 14C-labeled proteins were applied to a separate gradient to serve as standards.

of approximately 160kd. This band was more pronounced at even lower concentrations of SDS (0.05%). In addition, a faint 120kd band, also noted by Pinter et al. (31), was observed. Thus, under low SDS conditions, even monomeric gp41 migrated at an unexpectedly high MW. However, we note that only a fraction of the protein did so. Most continued to migrate in monomeric form, whereas material extracted from cells directly into low SDS without prior boiling and size fractionation migrated quantitatively at 160kd. This argues against the observed behavior of gp41 arising from anomolous migration secondary to reduced SDS binding at low detergent concentrations, in which case we would expect all of the monomeric gp41 to migrate at 160kd under low SDS conditions. Rather, the fact that only a fraction of monomeric gp41 migrated at 160kd (and some at 120kd) suggests that this arises from concentration dependent aggregation. Furthermore, since only discrete 160kd and 120kd bands were observed, the aggregation may be indicative of a native oligomeric interaction. While gp41 displays this interesting property, gp160 does not. Perhaps oligomeric gp41 is more stable to boiling in SDS which would, as discussed above, be consistent with our evidence that the ectodomain of gp41 is largely responsible for dimer formation and stability.

## DISCUSSION

The attainment of correct quaternary structure is an important post translational event for a large number of membrane and secretory proteins (15,16). In the vast majority of cases, assembly occurs in the ER and is aided by a number of resident proteins, including grp78 (BiP), protein disulfide isomerase, and proline isomerase.

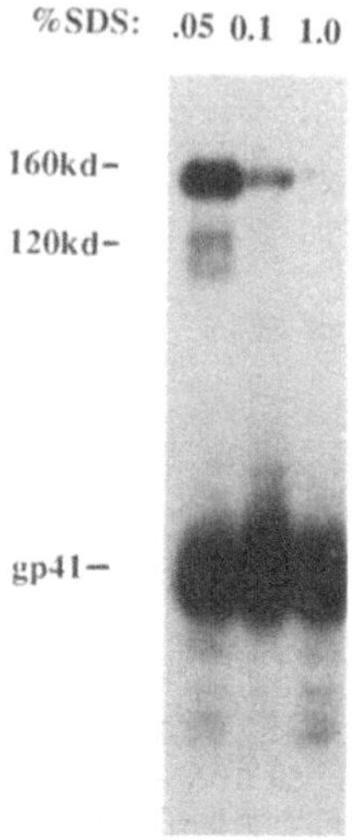


Fig. 8. Migration of monomeric gp41 under low SDS-conditions. Sucrose gradient purified monomeric gp41 (see figure 7) was boiled for 5min with 1mM DTT and the indicated concentration of SDS. Electrophoresis was done with 0.1% SDS. The distribution of gp41 was determined by western blotting.

When a protein misfolds due to a mutation, adverse cellular conditions (such as elevated temperature), or perhaps even by chance, it is usually retained in the ER where it is eventually degraded (15,16). Thus, correct assembly appears to be a prerequisite for export from the ER. Much of what is known about the relationships between protein folding, assembly, and transport are derived from studies on well characterized viral membrane proteins, with the influenza hemagglutinin and vesicular stomatitis virus G proteins being the most obvious examples (24,33-38). Indeed, it appears that most and perhaps all membrane proteins are oligomeric. While the quaternary structures of the orthomyxovirus (39), paramyxovirus (29,30), alphavirus (40), and rhabdovirus (24) membrane proteins are well characterized, much less is known about the structures of retrovirus env proteins. Previous reports suggested that the env protein of RSV is trimeric (32), while the env proteins of HIV-1, HIV-2 and SIV are monomers (27). However, Pinter et al. (31) and Schawaller et al. (48) have provided evidence which says the HIV-1 env protein is tetrameric. In addition, cross-linking studies on several other retroviral env proteins suggest that these too are oligomeric (41-43).

We have studied the oligomeric structure of the HIV-1 env protein using a variety of techniques in order to begin studies correlating its structure with function. We found that the env protein forms dimers shortly after synthesis. Dimers were detected by three independent assays: gradient centrifugation, resistance to SDS-induced dissociation, and chemical cross-linking. Given the agreement between these different approaches, there seems to be little doubt that the dimer represents a particularly stable form of the HIV-1 env protein. However, a higher order structure also exists. Specifically, we have detected a larger oligomer by both chemical cross-linking and gradient centrifugation with a MW of approximately 600kd and a sedimentation coefficient of 14-16S. The large size of this complex makes it difficult to ascertain its oligomeric structure with absolute certainty. First, well characterized MW markers of sufficiently large size are not available for SDS-PAGE. While the cross-linked ß-galactosidase used in our studies gave good results, we note that cross-linked proteins may migrate more quickly in SDS-gels than non cross-linked forms (26). This, coupled with the additional ambiguity imposed on the estimation of the env proteins's MW due to its high carbohydrate content makes determination of its absolute size even more difficult. Thus, the MW estimated for the higher order form may be consistent with the molecule being either trimeric (expected MW of 480kd) or tetrameric (expected MW of 640kd). Likewise, the sedimentation coefficient of 14S to 16S is between what would be expected of either trimeric or tetrameric env protein. However, the env protein is likely to sediment more slowly than expected for its size due to its elongated shape and detergent binding. Thus, $S_{20,w}$ values of from 14 to 16S are more likely to be consistent with tetrameric than trimeric molecules.

Several things lead us to conclude that the higher order form is tetrameric. First, recovery of stable dimers suggests that these form the "building blocks" of any higher order structure. Thus, a tetramer composed of two dimers or a hexamer composed of three dimers would be allowable structures, but a trimer composed of a dimer and a monomer would not. Such an asymmetrical arrangement has not been observed for other homo-oligomers (15,16). Second, the estimations of MW and sedimentation coefficient are closer to what would be expected for a tetrameric rather than a trimeric molecule. Third, the env proteins of paramyxoviruses, with which the HIV-1 env protein shares some structural and functional homology, are themselves tetramers composed of two dimers (29,30,49). Finally, we have sometimes observed a cross-linked protein with MW and $S_{20,w}$ values intermediate between the dimeric 10.8S form of the molecule and the putative 14S-16S tetramer. Obviously, this is likely to represent the trimeric species. A subunit structure of $A_2A_2$ would easily explain our inability to observe this partially cross-linked intermediate to any great degree. Thus, given these considerations, we concluded that the env protein forms dimers which in turn assemble into tetramers composed of two dimers each. Important questions which remain to be addressed include where in the cell tetramers are formed, with what efficiency tetramers are formed, and whether both dimers and tetramers are fully functional entities that can be transported to the cell surface and incorporated into virus.

The structure of an env protein dimer is represented schematically in Figure 9. We depict the gp120 subunits as constituting discrete, compact, globular domains for several reasons. For example, gp120 expressed in the absence of gp41 folds into a conformation competent for CD4 binding and is secreted efficiently (10). Since a large amount of gp120 is eventually shed from the cell surface, interactions between gp120 and gp41 are apt to be relatively weak (44). We have also not yet detected direct interactions between adjoining gp120 molecules in the env protein though they may well exist. Since gp120 does not oligomerize in the absence of gp41 and we have not observed cross-linking of adjoining gp120 molecules, we show adjacent gp120 domains as having little direct contact. By analyzing the properties of a series of truncated forms of the env protein, we were able to show that the amino terminal 129 residues of gp41, comprising most of the molecule's

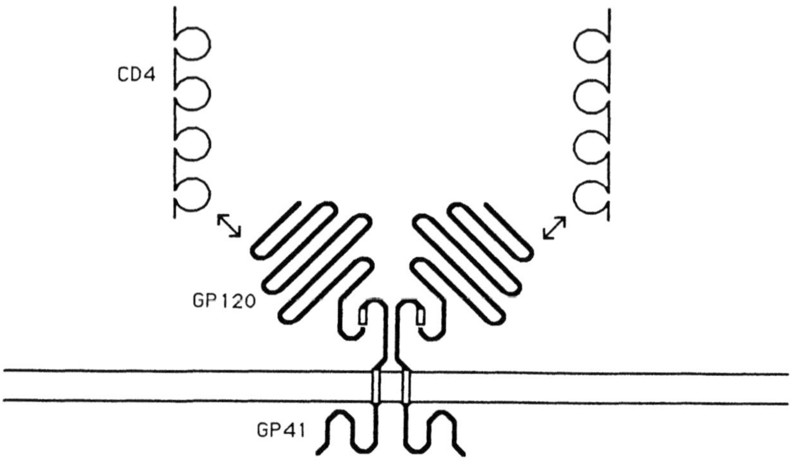


Fig. 9. Model for the structure of an HIV-1 env protein dimer.

ectodomain, are both sufficient and necessary for dimer formation and stability. Thus, we depict the ectodomains of gp41 as interacting with one another closely. While the transmembrane and cytoplasmic domains may well interact with one another, these interactions are not likely to be as intimate as those in the ectodomain region. Characterization of the env protein's oligomeric structure should help direct efforts to obtain water soluble oligomers for more detailed structural studies.

## ACKNOWLEDGEMENTS

We would like to thank our colleagues Dr. E. Berger and Dr. S. Chakrabarti for helpful discussions throughout the course of this work. Figures 1 through 5 have appeared in P.L. Earl, R.W. Doms and B. Moss, Oligomeric structure of the human immunodeficiency virus type 1 envelope glycoprotein, PNAS, in press, and are reprinted here with permission.

## REFERENCES

1. A.G. Dalgleish, P.C.L. Beverley, P.R. Clapham, D.H. Crawford, M.F. Greaves, and R.A. Weiss, The CD4 (T4) antigen is an essential component of the receptor of the AIDS retrovirus, Nature 312:763 (1984).
2. D. Klatzman, E. Champagne, S. Chamaret, J. Gruest, D. Guetard, T. Hercend, J.C. Gluckman, and L. Montagnier, T-lymphocyte T4 molecule behaves as the receptor for human retrovirus LAV, Nature 312:767 (1984).
3. J.S. McDougal, A. Mawie, S. Cort, J. Nicholson, G.D. Cross, J.A. Scheppler-Campbell, D. Hicks, and J. Sligh, Cellular tropism of the human retrovirus HTLVIII/LAV. 1. Role of T-cell activation and expression of the T4 antigen, J. Immunol. 135:3151 (1985).

4. J.S. McDougal, M.S. Kennedy, J.M. Sligh, S.P. Cort, A. Mawle, and J.K.A. Nicholson, Binding of HTLV-III/LAV to T4+ T cells by a complex of the 110K viral protein and the T4 molecule, Science 231:383 (1986).
5. P.J. Maddon, A.G. Dalgleish, J.S. McDougal, P.R. Clapham, R.A. Weiss, and R. Axel, The T4 gene encodes the AIDS virus receptor and is expressed in the immune system and the brain, Cell 47:333 (1986).
6. B.S. Stein, S.D. Gowda, J.D. Lifson, R.C. Penhallow, K.G. Bensch and E.G. Engelman, pH-independent HIV entry into CD4-positive T cells via virus envelope fusion to the plasma membrane, Cell 49:659 (1987).
7. M. Kowalski, J. Potz, L. Basiripour, T. Dorfman, W.C. Goh, E. Terwillger, A. Dayton, C. Rosen, W. Haseltine and J. Sodroski, Functional regions of the envelope glycoprotein of human immunodeficiency virus type I, Science 237:1351 (1987).
8. J.S. Allan, J.E. Coligan, F. Barin, M.F. McLane, J.G. Sodroski, C.A. Rosen, W.A. Haseltine, W.A. Lee, and M. Essex, Major glycoprotein antigens that induce antibodies in AIDS patients are encoded by HTLV-III, Science 228:1091 (1985).
9. F.D. Veornese, A.L. De Vico, T.D. Copeland, S. Oroszlan, R.C. Gallo, and M.G. Sarngadharan, Characterization of gp41 as the transmembreane protein coded by the HTLV-III/LAV envelope gene, Science 229:1402 (1985).
10. C. Fennie and L. Lasky, Model for intracellular folding of human immunodeficiency virus type 1 gp120, J. Virol. 63:639 (1989).
11. R.L. Willey, J.S. Bonifacino, B.J. Potts, M.A. Martin and R.D. Klausner, Biosynthesis, cleavage and degradation of the human immunodeficiency virus 1 envelope glycoprotein gp160, Proc. Natl. Acad. Sci. 85:9580 (1988).
12. J.M. McCune, L.B. Rabin, M.B. Feinberg, M. Lieberman, J.C. Kosek, G.R. Reyes, and I.L. Weissman, Endoproteolytic cleavage of gp160 is required for the activation of human immunodeficiency virus, Cell 53:55 (1988).
13. J. White, M. Kielian, and A. Helenius, Membrane fusion proteins of enveloped animal viruses, Q. Rev. Biophys. 16:151 (1983).
14. T. Stegmann, R.W. Doms, and A. Helenius, Protein-mediated membrane fusion, Annu. Rev. Biophys. Chem. 18:187 (1989).
15. J.K.Rose and R.W. Doms, Regulation of protein export from the endoplasmic reticulum, Ann. Rev. Cell Biol. 4:257 (1988).
16. S.M. Hurtley and A. Helenius, Protein oligomerization in the endoplasmic reticulum, Annu. Rev. Cell Biol., in press.
17. J. Lippincott-Schwartz, J.S Bonifacino, L.C. Yuan and R.D. Klausner, Degradation from the endoplasmic reticulum: Disposing of newly synthesized proteins, Cell 54:209 (1988).
18. L. Ratner, W. Haseltine, R. Patarca, K.J. Livak, B. Starchch, S.F. Josephs, E.R. Doran, J.A. Rafalski, E.A. Whitehorn, K. Baumesiter, L. Ivanoff, S.R. Petteway, M.L. Pearson, J.A. L.J. Takemoto, C.F. Fox, F.C. Jensen, J.H. Elder, and R.A. Lerner, Nearest-neighbor interactions of the major RNA tumor virus glycoprotein on cell surfaces, Proc. Natl. Acad. Sci. 75:3644 (1978).
19. S. Chakrabarti, M. Robert-Guroff, F. Wong-Stahl, R.C. Gallo, and B. Moss, Expression of the HTLV-III envelope gene by a recombinant vaccinia virus, Nature 320:535 (1986).
20. P. Earl and B. Moss, in preparation.

21. C.R. McEwen, Tables for estimating sedimentation through linear concentration gradients of sucrose solution, Anal. Biochem. 20:114 (1967).
22. R.G.Martin and B.N. Ames, A method for determining the sedimentation behavior of enzymes: application to protein mixtures. J. Biol. Chem. 236:1372 (1961).
23. R.W.Doms and A. Helenius, Quaternary structure of influenza virus hemagglutnin after acid treatment, J. Virol. 60:833 (1986).
24. R.W. Doms, D.S. Keller, A. Helenius and W.E. Balch, Role for ATP in regulating the assembly and transport of Vesicular Stomatitis virus G protein trimers, J. Cell Biol. 105:1957 (1987).
25. J.D. Lifson, M.B. Feinberg, G.R. Reyes, L. Babins, B. Bonapour, S. Chakrabarti, B. Moss, F. Wong-Stahl, D.C. Steiner, and E.G. Engelman, Induction of CD4-dependent cell fusion by the HTLV-III/LAV envelope glycoprotein, Nature 323:725 (1987).
26. R.W. Doms, Oligomerization and protein transport, Meth. Enzymology, in press.
27. M.-A. Rey, B. Krust, A.G. Laurent, L. Montagnier and A.G. Hovanessian, Characterization of human immunodeficiency virus type 2 envelope glycoproteins: Dimerization of the glycoprotein precursor during processing, J. Virol. 63:647 (1989).
28. M. Silverberg and V.T. Marchesi, The anomolous electrophoretic behavior of the major sialoglycoprotein from the human erythrocyte, J. Biol. Chem. 253:95 (1978).
29. O. Sechoy, J.R. Philippot, and A. Bienvenue, F protein-F protein interaction within the Sendai virus identified by native bonding or chemical cross linking, J. Biol. Chem. 262:11519 (1987).
30. M.N. Waxham, D.C. Merz, and J.S. Wolinsky, Intracellular maturation of mumps virus hemagglutinin-neuraminidase glycoprotein: Conformational changes detected with monoclonal antibodies, J. Virol. 59:1186 (1986).
31. A. Pinter, W.J. Honnen, S.A. Tilley, C. Bona, H. Zaghouani, M.K. Gorny, and S. Zolla-Pazner, Oligomeric structure of gp41, the transmembrane protein of human immunodeficiency virus type 1, J. Virol. 63:2674 (1989).
32. D. Einfeld and E. Hunter, Oligomeric structure of a prototype retrovirus glycoprotein, Proc. Natl. Acad. Sci. 85:8688 (1988).
33. R.W. Doms, A. Ruusala, C. Machamer, J. Helenius, A. Helenius and J.K. Rose, Differential effects of mutations in three domains on folding, quaternary structure, and intracellular transport of VSV G protein, J. Cell Biol. 107:89 (1988).
34. C.E. Machamer, and J.K. Rose. Vesicular stomatitis virus G proteins with altered glycosylation sites display temperature-sensitive intracellular transport and are subject to aberrant intermolecular disulfide bonding, J. Biol. Chem. 263:5955 (1988).
35. C.S. Copeland, K.-P. Zimmer, K.R. Wagner, G.A. Healey, I. Mellman and A. Helenius, Folding, trimerization and transport are sequential events in the biogenesis of influenza virus hemagglutinin, Cell 53:197 (1988).
36. C.S. Copeland, R.W. Doms, E.M. Bolzau, R.G. Webster and A. Helenius, Assembly of influenza hemagglutinin trimers and its role in intracellular transport, J. Cell Biol. 103:1179 (1986).
37. M.-J. Gething, K. McCammon, and J. Sambrook, Expression of wild-type and mutant forms of influenza hemagglutinin: The role of folding in intracellular

transport, Cell 107:939 (1986).
38. P. Gallagher, J. Henneberry, I. Wilson, J. Sambrook, and M.-J. Gething, Addition of carbohydrate side-chains at novel sites on influenza virus hemagglutinin can modulate the folding, transport, and activity of the molecule, J. Cell Biol. 107:2059 (1988).
39. I. Wilson, J. Skehel, and D.C. Wiley, Structure of the hemagglutinin membrane glycoprotein of influenza virus at 3Å resolution, Nature 292:366 (1981).
40. R.H. Vogel, S.W. Provencher, C.-M. von Bonsdorff, M. Adrian, and J. Dubochet, Envelope structure of Semliki Forest virus reconstituted from cryoelectron micrographs, Nature 320:533 (1986).
41. L.J. Takemoto, C.F. Fox, F.C. Jensen, J.H. Elder, and R.A. Lerner, Nearest-neighbor interactions of the major RNA tumor virus glycoprotein on cell surfaces, Proc. Natl. Acad. Sci. 75:3644 (1978).
42. A. Pinter and E. Fleissner, Structural studies of retroviruses: Characterization of oligomeric complexes of murine and feline leukemia virus envelope and core components formed upon cross-linking, J. Virol. 30:157 (1979).
43. J. Racevskis and N.H. Sarkar, Murine mammary tumor virus stuctural protein interactions: Formation of oligomeric complexes with cleavable cross-linking agents, J. Virol. 35:937 (1980).
44. H.R. Gelderbloom, H. Reupke, and G. Pauli, Loss of envelope antigens of HTLV-III/LAV, a factor in AIDS pathogenesis? Lancet ii:1016 (1985).
45. B. Chesebro and K. Wehrly, Development of a sensitive quantitative focal assay for human immunodeficiency virus infectivity, J. Virol. 62:3779 (1988).
46. A. Helenius and K. Simons, Solubilization of membranes by detergents, Biochim. Biophys. Acta 415:29 (1975).
47. C. Tanford and J.A. Reynolds, Characterization of membrane proteins in detergent solutions, Biochim. Biophys. Acta, 457:133 (1976).
48. M. Schawaller, G.E. Smith, J.J. Skehel, and D.C. Wiley, Studies with crosslinking reagents on the oligomeric structure of the env glycoprotein of HIV, Virology 172:367 (1989).
49. M.A.K. Markwell and C.F. Fox, Protein-protein interactions within paramyxoviruses identified by native disulfide bonding or reversible chemical cross-linking, J. Virol. 33:152 (1980).

## DISCUSSION

Larsen (UCSF): On the gel electrophoresis, what molecular weight standards did you use to distinguish between 320, 480 and 640 kD proteins?

Doms: We run the material on 4% acrylamide gels and use cross-linked phosphorylase B as a molecular weight standard. More recently, we have used cross-linked B-galactosidase as a standard. Obviously, assigning an accurate molecular weight to a very large, heavily glycosylated cross-linked protein, sometimes under non-reducing conditions, is difficult and in the final analysis, several approaches will have to be taken to unambiguously determine the final oligometric structure of the HIV-1 env protein.

Larsen: Is there a clear difference at the top of the gel between the different oligomers of the phosphorylase?

Doms: In large, low percentage acrylamide gels, yes.

Larsen: And one last question. When you're comparing the cross-linked proteins, non-reduced and reduced, you're running them on a gel that's non-reduced?

Doms: Reduced and non-reduced samples are analyzed on identical, but separate gels.

Levy (UCSF): First, let me say it was a very nice presentation and a lot of work. I wonder whether you could form the criticism that I couldn't give, which would be that you're using an artificial model. You're using vaccinia-expressed envelope. Can we assume that the virus does the same or must we then go and actually look at the virus? Obviously the question would be, have you done that?

Doms: Yes. I think it is reasonable to conclude that the vaccinia system accurately reflects the situation in native virus. First, we've looked at 8E5 cells and the sedimentation patterns are the same. Second, Schawaller et al., in a recent paper, have reached conclusions basically identical to our own. They did do some analyses on intact HIV. Finally, numerous viral membrane proteins have been expressed with the vaccinia system. In all of the cases I am aware of, the proteins fold and oligomerize normally.

Levy: An educated guess may be based on the paramyxovirus story. We have spikes on the virus. Are those spikes dimers, tetramers, or are they . . .

Doms: The paramyxovirus envelope proteins, F and HN, are believed to be tetramers. In both instances, the actually seem to consist of a dimer of dimers. This is the same structural motif we are proposing for the HIV-1 env protein.

Engleman (Stanford): Could you say what your evidence is that shows the gp160 is actually expressed on the surface? And do you know of any evidence in HIV-infected cells that gp160 is expressed on the surface?

Doms: At present, we do not have a reliable quantitative assay to measure delivery of env protein to the cell surface. I should note that we have a processing mutant of the env protein which is not cleaved to gp120/gp41. This protein is expressed on the cell survace judging by immunofluorescence. As for HIV-infected cells, I do not know if gp160 is ever expressed on the cell surface. We have not done those experiments.

Engleman: I don't think there's any dispute about the process mutant. The question is whether or not it actually happens in HIV-infected cells. Did you try to iodine-label any of this material?

Doms: No, we have not yet tried to iodinate cell surface env protein.

Stein (Stanford): I'd like to address the whole question of gp160 on the cell surface. I think you have to be very careful about the interpretation. In our lab we have used de novo infected $CD4^+$ T cells, and in that particular system we don't see any gp160 on the cell surface. I think we need to look at the literature. There was some work done by Eric Hunter in the RSV system where they showed that all you need is a one amino acid change in the cleavage region from a lysine to glutamic acid and the precursor of the RSV envelope, which is not normally expressed on the cell surface, then gets expressed on the cell surface. So I think that when we start making constructs, and we start moving around nucleotide sequences, that small changes in retention and sorting signals within the protein can make vast differences in how it's handled through the Golgi complex. Our experience is that in our system all our gp160 is endo-H sensitive. Now, we also know that the glycosylation between gp120 and gp160 is quite variable.

CONTRIBUTORS

Dennis R. Alford: Cancer Research Institute, University of California, San Francisco, CA 94143-0128. Present address: Department of Bioscience and Biotechnology, Drexel University, Philadelphia, PA 19104.

R. A. Axel: Department of Biochemistry and Molecular Biophysics and the Howard Hughes Medical Institute, College of Physicians and Surgeons, Columbia University, New York, NY 10032.

P. R. Clapham: Institute of Cancer Research, Fulham Road, Chester Beatty Laboratories, London, SW3 6JB, England.

A. Cordonnier: Unité d'Oncologie Virale (CNRS UA 1157), Institut Pasteur, 75724 Paris Cedex 15, France.

Suzanne Crowe: MacFarlane Burnet Centre for Medical Research, Fairfield, Victoria, Australia.

A. G. Dalgleish: Retrovirus Research Group, Division of Immunology, Clinical Research Centre, Northwick Park Hospital, Watford Road, Harrow HA1 3UJ, United Kingdom.

Brian R. Davis: Medical Research Institute of San Francisco at Pacific Presbyterian Medical Center, San Francisco, CA 94115.

Robert W. Doms: Laboratory of Viral Diseases, National Institute of Allergy and Infectious Diseases, and Laboratory of Pathology, National Cancer Institute, National Institutes of Health, Bethesda, MD 20892.

Nejat Düzgünes, Cancer Research Institute and Department of Pharmaceutical Chemistry, University of California, San Francisco, CA 94143-0128; and Department of Microbiology, University of the Pacific School of Dentistry, San Francisco, CA 94115.

Patricia L. Earl: Laboratory of Viral Diseases, National Institute of Allergy and Infectious Diseases, National Institutes of Health, Bethesda, MD 20892.

Edgar Engleman: Stanford Blood Center, Stanford University School of Medicine, Palo Alto, CA 94305.

Suzanne Gartner: Primate Research Institute, New Mexico State University, P.O. Box 1027, Holloman AFB, New Mexico 88330.

William Haseltine: Laboratory of Human Retrovirology, Department of Pathology, Harvard Medical School, Dana-Farber Cancer Institute, and Department of Cancer Biology, Harvard School of Public Health, Boston, MA 02115

James A. Hoxie: Hematology-Oncology Section, Hospital of the University of Pennsylvania, 3400 Spruce Street, Philadelphia, PA 19104.

S. Jamal: Department of Biochemistry and Molecular Biophysics and the Howard Hughes Medical Institute, College of Physicians and Surgeons, Columbia University, New York, NY 10032.

Myra Jennings: Comparative Oncology Laboratory, University of California, Davis, CA 95616.

Krystyna Konopka, Cancer Research Institute, University of California, San Francisco, CA 94143-0128. Present address: Department of Microbiology, University of the Pacific School of Dentistry, San Francisco, CA 94115.

Mark Kowalski: Laboratory of Human Retrovirology, Department of Pathology, Harvard Medical School, Dana-Farber Cancer Institute, Boston, MA 02115

Charles E. Larsen: Cancer Research Institute, University of California, San Francisco, CA 94143-0128. Present address: Department of Bioscience and Biotechnology, Drexel University, Philadelphia, PA 19104.

Jay A. Levy: Department of Medicine and Cancer Research Institute, University of California, San Francisco, CA 94143-0128.

P. J. Maddon: Department of Biochemistry and Molecular Biophysics and the Howard Hughes Medical Institute, College of Physicians and Surgeons, Columbia University, New York, NY 10032.

J. S. McDougal: Immunology Branch, Division of Host Factors, Center for Infectious Diseases, Centers for Disease Control, Atlanta, GA 30333.

Michael McGrath: Medical Service, San Francisco General Hospital, and Departments of Medicine and Laboratory Medicine, University of California, San Francisco, CA 94143.

Thomas P. McGraw: California Primate Research Center, University of California, Davis, CA 95616.

John Mills: Medical Service, San Francisco General Hospital, and Departments of Medicine, Microbiology and Laboratory Medicine, University of California, San Francisco, CA 94143.

L. Montagnier: Unité d'Oncologie Virale (CNRS UA 1157), Institut Pasteur, 75724 Paris Cedex 15, France.

Bernard Moss: Laboratory of Viral Diseases, National Institute of Allergy and Infectious Diseases, National Institutes of Health, Bethesda, MD 20892.

Jay A. Nelson: Department of Immunology, Research Institute of Scripps Clinic, and Department of Pathology, University of California San Diego, La Jolla, CA.

Shlomo Nir: Department of Soil and Water Sciences, The Hebrew University of Jerusalem, Rehovot 76100, Israel, and Cancer Research Institute, University of California, San Francisco, CA 94143-0128.

Kunihiro Ohashi: Primate Research Institute, New Mexico State University, P.O. Box 1027, Holloman AFB, New Mexico 88330.

G. Orloff: Immunology Branch, Division of Host Factors, Center for Infectious Diseases, Centers for Disease Control, Atlanta, GA 30333.

C. David Pauza: Department of Pathology and Laboratory Medicine, University of Wisconsin-Madison Medical School, 505 Services Memorial Institute, 1300 University Avenue, Madison, WI 53706.

Mikulas Popovic: Primate Research Institute, New Mexico State University, P.O. Box 1027, Holloman AFB, New Mexico 88330.

Catherine Reynolds-Kohler: Department of Immunology, Research Institute of Scripps Clinic, and Department of Pathology, University of California San Diego, La Jolla, CA.

Joseph Sodroski: Laboratory of Human Retrovirology, Department of Pathology, Harvard Medical School, Dana-Farber Cancer Institute, Boston, MA 02115

Barry Stein: Stanford Blood Center, Stanford University School of Medicine, Palo Alto, CA 94305.

R. A. Weiss: Institute of Cancer Research, Fulham Road, Chester Beatty Laboratories, London, SW3 6JB, England.

Clayton Wiley: Department of Immunology, Research Institute of Scripps Clinic, and Department of Pathology, University of California San Diego, La Jolla, CA.

Lawrence J. T. Young: California Primate Research Center, University of California, Davis, CA 95616. Present address: Department of Pathology, University of California, Davis, Medical Center, Sacramento, CA 95817

INDEX

**Batch number: 10684222**

Printed by Printforce, the Netherlands